ENERGY 2000

An Overview of the World's Energy Resources in the Decades to Come

Heinz Knoepfel
Association EURATOM-ENEA
Energy Research Center
Frascati (Rome)

GORDON AND BREACH SCIENCE PUBLISHERS
New York • London • Paris • Montreux • Tokyo

Gordon and Breach Science Publishers

P.O. Box 161
1820 Montreux 2
Switzerland

P.O. Box 786
Cooper Station
New York, NY 10276
United States of America

P.O. Box 197
London WC2E 9PX
England

58, rue Lhomond
75005 Paris
France

14-9 Okubo 3-chome,
Shinjuku-ku,
Tokyo 160
Japan

Library of Congress Cataloging-in-Publication Data

Knoepfel, Heinz, 1931—
    Energy 2000.

    Bibliography: p.
    Includes index
    1. Power resources. 2. Power (Mechanics) I. Title.
TJ163.2.K577   1986   333.79   86—3145
ISBN 2-88124-074-7

Contents

CONTENTS

Preface

The aim of this book is to present a comprehensive, if succinct, picture of energy, probably the most significant single quantity defining the world in and around us. Primary consideration is given to topics relevant to the medium to long term, and to the elementary explanation of technical concepts such that an attentive but not necessarily scientifically-trained reader would understand. In addition, one chapter and frequent flashbacks provide a historical perspective which should help when pondering about the many interrelated and interdisciplinary aspects of the energy problem.

By setting the focus on the future, the intricate energy issue frees itself from present-day economic and political constraints, and becomes fascinating and stimulating. From this point of view, even the often dramatically debated nuclear energy—to which this book pays particular attention—becomes an interesting and challenging topic for discussion. Since the potential energy resources on planet Earth are more than plentiful, the future choices among the many technical energy options can be made by amply taking into account ecological, human, and ethical considerations, rather than only economic convenience.

The extension of the main subject demanded selection and often indepth analysis of the topics to be treated, so as to allow a meaningful and synthetic presentation. Admittedly, this is one place where a certain subjectiveness could have sneaked into the book. If this has happened, in spite of my experience on the energy problem gained during nearly 30 years of scientific research in this field, I have to apologize.

This book grew out of my personal involvement, outside my research activity, in communicating with people on the subject of energy. In the winter of 1976–1977 I held a series of six television programs on the Swiss Television Network, which were presented during prime viewing time. I later used my TV manuscript as the basis for a book in Italian, *Il problema energia*, published by Franco Angeli, Milan. After long and stimulating discussions with students and visitors to our Frascati Research Center—taking careful note of their interest, biases, lack of information on the subject, and above all their enthusiasm for learning and debating about it—I decided to write this book.

In view of the general nature of the book, the quoting of bibliographical sources has most often been avoided, unless the data or statements are very source-related. On the other hand, I felt it useful to mention some general literature for further reading.

I am indebted to many colleagues, both in the United States and in Europe—too numerous to mention—who made useful suggestions for the improvement of the text. Obviously, the views expressed in the book are the author's personal ones and are not intended to represent those of the Commission of the European Community (to whom I belong) or of the Italian Commission for Nuclear and Alternative Energy Sources, ENEA (where I work).

On a more personal note, I wish to express my particular appreciation to Lucilla Crescentini for her skillful and dedicated editorial handling of the several versions of my manuscript, and to Peter Riske for the careful preparation of the drawings. I am also grateful to Maria Polidoro for her efficient and competent secretarial assistance in all phases of my work.

Heinz Knoepfel

CHAPTER 1

Energy Today

Energy is perhaps the most significant single quantity defining the world in and around us. It is of basic importance in physics and biology, as well as in most manifestations of our industrialized society. Energy means not only fuel for transportation and for heating, fertilizers for agriculture, and chemical products for industry, but also food and housing for man's well-being. The great scientist Boltzmann suggested in 1886 that life is primarily a struggle for energy.

1.1 AN ENERGY PRIMER

Energy has always played an explicit and important role in the development of our civilization, particularly so in the last two centuries. The advent of the steam engine at the end of the eighteenth century heralded a new and fundamental application of energy which was to provide the motive force for the socioeconomic revolution of our modern era. One of the immediate consequences was to shift the textile mills from the countryside (where they harnessed power from watercourses) to the cities. The birth of industry came *Energetic* about in socially difficult circumstances; not until a century or so *unrest* later were the working masses to share to any real extent the benefits of the greater productivity made possible by this new form of energy.

Another sector revolutionized by the steam engine was transport. The year 1825 saw the inauguration in England of the first railway in *Steamy* history. On the basis of the great number of empirical results *railway* obtained by the manufacturers of steam engines, a more precise and scientific analytical approach was developed, which eventually provided the basis of thermodynamics and, on a more general level, of modern physics.

Energy is today only partially a technological problem, as it depends strongly also on socioeconomic and political considerations and, above all, on fundamental choices to be made on the basis of moral value judgments on our way of life.

In 1973, when the effects of the hard line adopted by the major Arab oil producers in the wake of events in the Middle East reached the west, it was almost as if a fairground had been hit by a power cut. Nevertheless, at first through inertia and then out of sheer *Lights at* necessity, the big wheel of the consumer society did not in fact stop *the* turning, although it did lose some of its momentum temporarily and *fairground* there were economic difficulties in certain areas. The system has rapidly adapted itself to the new situation, and energy on the market is again plentiful. Those years already seem far away and the preoccupations about energy forgotten.

Nevertheless, what has remained is the awareness of people of the almost total dependence of our way of life on what could in fact be defined as the most important raw material of the world economy, i.e., energy in its many diverse forms. The 1973 crisis also revealed the vast economic interests and political implications at international *Energy in* levels which center on energy. At any rate, it heralded the end of a *politics* consumer society based on very cheap energy which had been the hallmark of the 1960s, and the beginning of a positive, although not altogether painless development toward a new economic situation characterized by what could be described as a slowed-down consumer society. For the rich, industrialized nations the containment of energy consumption turned out to be a positive process. The situation was, and is, totally different in the developing and poor countries where progress is conditioned upon a substantial increase in the consumption of energy, i.e., on the necessity to have cheap and abundant energy.

The economic implications of energy are impressive and far-reaching. Apart from the final commercial cost of energy, something of which we, as consumers, are perfectly aware, the huge investment needed to exploit energy sources must also be considered. At the eleventh World Energy Conference in Munich in 1980, a report by the Dresdner Bank estimated that the non-Communist countries will have to spend up to $10 trillion ($10^{13}$ in constant 1979 dollars) to *Demanding* meet the energy demand up to the year 2000. This enormous capital *investments* investment would be required to restructure present energy supplies (e.g., oil will come from more remote and hostile areas, and will require increasingly sophisticated recovering techniques); to shift the emphasis away from oil more toward nuclear energy, gas and coal; to increase the share of electricity in the overall energy consumption; to build the necessary gross distribution systems (pipelines and electricity grids). Even if the effective capital requirement should turn out to be much less than anticipated, as a consequence of changing energy scenarios and lower consumption levels (see in *Dollars in* Sec. 5.1), the mentioned figure stresses the importance of the *a barrel* energy business in the investment market.

The exploitation of a new energy source generally requires huge extracting facilities and power plants, which in themselves absorb large amounts of energy to be set up: The preparation of materials, such as the extraction and processing of metals, requires energy as well as raw materials. Consider, for a nuclear source, the energy required for the whole uranium fuel cycle (from ore extraction and *Energy* purification to enrichment and preparation of the reactor fuel, up to *payoff* the reprocessing of the spent fuel; see Chap. 3) and the massive power station itself. The energy payoff time in this case is typically 3 to 5 years, meaning that the energy output of the electronuclear power plant over 3 to 4 years is needed to cover the energy investment required to get the plant into operation in the first place.

This interconnection between energy and materials also points to a critical dependence of the cost of materials (and of food) on the cost of energy; and vice versa (Sec. 1.2 and 1.3). This is one of the reasons why the poor and developing countries (without their own energy sources) must be most concerned about the future avail- *The oil* ability and cost of energy, whereas the industrialized western *eater* nations should well be able to ride the wave of energy-related problems and manage the corresponding financial difficulties.

Large consumption of energy by man negatively influences the environment, a problem that mostly hits the consumer society. In large conventional power stations, as we shall see, roughly one-third of the primary energy derived from oil, coal, or nuclear sources is converted into electricity; the remaining two-thirds are released as heat into the atmosphere surrounding the power station, whence the so-called problem of thermal pollution. In order not to disrupt the *Heat not* ecological equilibrium by excessive heating of natural waters, cool- *wanted* ing towers are often needed and these constitute the most imposing and sight-disturbing part of the installation.

A more serious problem is atmospheric pollution caused by certain types of conventional power stations. A large, but standard 1 gigawatt (1 billion watts) coal-fired electric power station consumes 10 000 tons of coal daily, much of which is released into the atmosphere in the form of carbon dioxide and also produces roughly 600 tons of ash and over 200 tons of sulfur dioxide. Despite the filter required by the new laws in force, a significant part of these pro-

Smoke gets in the air ducts is released into the atmosphere and is then dispersed over a large area.

Nuclear energy could be one of the important primary energy sources of the future (Sec. 4.5). Although it has already an important share (e.g. in 1984 one quarter of the European Community's electricity is from nuclear origin), its future role is still open to debate (Sec. 3.1 and 3.3).

An underlying feature of these various aspects of the energy problem is poverty: 25% of the world's population suffers from malnutrition because of poverty, and approximately half a million

No learned debates human beings die of hunger every year. Poverty is a type of pollution which generates suffering and hate and which kills. And poverty cannot be overcome unless enough energy is available. The poor peoples of the world are not overly concerned by our learned debates on pollution and the potential dangers of radioactivity; these drawbacks can be eliminated or mitigated by means of suitable, possibly expensive measures which will in any case have to be paid for by the richer nations.

Many of the problems discussed or mentioned so far in this chapter find a logical ordering around the evolution curve shown in

Riding on the top Fig. 1.1. As energy consumption per capita increases, income also increases in a spiraling interconnection, and the nations thereby develop from preindustrial into an industrial society. This is the transformation in which most countries on the earth are presently involved—some are still far behind; others, like the industrialized western nations, ride on the top of this evolution.

In relation to the industrial evolution, it is customary to define the following industrial capabilities or stages: primary industries, such as resource extraction or agriculture; secondary or manufacturing industries; tertiary service industries; and quaternary industries

A lean mix based on information treatment. An industrial society will have a mixture of all four types, with the tertiary and quaternary stages prevailing more and more as it evolves toward the postindustrial phase. These two stages increase the product value mainly through scientific and technical knowledge (i.e., "technology") and through

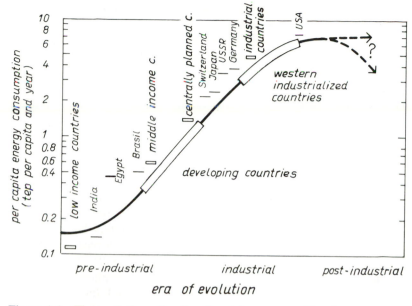

Figure 1.1   The evolution of national entities measured by the normalized energy consumption, in tons equivalent petroleum (tep) per capita and year (approx. 1982; notice the logarithmic scale of the consumption)

an efficient service infrastructure, and thus are relatively modest energy consumers. An interesting energy-related question is: How much will the energy consumption change as the societies evolve further into the postindustrial phase?

Many analysts think that the per capita energy needs can be reduced substantially with respect to the present consumption levels of some industrialized countries, without loosing anything in the quality of life. Supporting this belief is the fact that the Federal Republic of Germany and Switzerland consume one-half of the *Less is* energy used per capita in the United States, yet the standard of *beautiful* living is roughly the same. Even after due allowance has been made for the different economic and industrial structures of the three countries, this example sustains the belief that the industrialized west could, if it has a mind to and sufficient interest, substantially reduce its average consumption of energy. The leveling off, and even reduction, of the energy consumption since about 1979 may indicate that this evolution has already started.

The energy problem will continue to determine in a critical way most aspects of our present and future world, even of our very *World's* existence. In recent years, it has been analyzed and debated in *troubles* depth by the general public as well as by specialists, and it has been mentioned together with the other crucial and challenging issues which will face human society in the decades and centuries to come, such as:

- Nutrition of the hungry
- Conservation of the earth's ecosystem
- Control of sociopolitical unrest deriving from the uneven distribution among the nations or areas of this world of food, wealth, and opportunities, or just from the political culture of societies
- Prevention of global nuclear war
- Cure of psychosocial tension deriving from unemployment, particularly of the young generation; or from the numerical explosion of old people, this being a consequence of the huge medical research enterprise of the last 30 years
- Introduction of new technologies into society and their consequences, for example, microelectronics, robotics, new communications and information handling, biotechnology (including genetic engineering), new materials, exploitation of the oceans and of outer space

The energy problem is, however, intrinsically different from these issues, being from a materialistic point of view (as the following sections and chapters of this book want to show) both more fundamental and, at the same time, less dramatic. In fact, some of the mentioned issues relate critically to energy, in the sense that the availability of abundant energy at convenient costs is a necessary (but not sufficient) condition for tackling their solutions. Fortunately, it also is a less dramatic issue, since the energy available on earth is, *Energy, no* after all, plentiful, and today it is practicable to exploit it—this *problem* because the necessary huge investment capability and sophisticated technology are at hand (having been accumulated over the last centuries), or realistically foreseeable.

1.2 ENERGY FOR PEDESTRIANS

Definitions and equivalences

The word energy, derived from Greek (*en* = in, *ergon* = work) and used by Aristoteles, stands for one of the fundamental *Aristoteles'* quantities of physics which was given a precise definition, however, *work* only with the development of thermodynamics in the last century. For general purposes suffice it to define energy as a quantity with the capacity of producing work and/or heat.

Energy is measured in various units, the main one being the joule (symbol J; see Table 1.1). Another is the calorie, which in the past was commonly used to measure heat (thermal energy); it is defined as the thermal energy required to raise the temperature of 1 gram of water by 1°C (one degree Celsius), and the relation between the two is 1 calorie = 4.184 joules. Power, on the other hand, is defined as

Table 1.1   Energy units

1 kWh	$= 860 \times 10^3$ cal $= 3.6 \times 10^6$ J
1 cal	$= 1.161 \ 10^{-6}$ kWh $= 4.184$ J
1 erg	$= 10^{-7}$ J
1 eV	$= 1.602 \times 10^{-19}$ J
1 Btu	$= 1.055 \times 10^3$ J
1 quad	$= 10^{15}$ Btu $= 1.055 \times 10^{18}$ J
1 Q	$= 10^{18}$ Btu $= 1.055 \times 10^{21}$ J
1 bbl	$= 6.1 \times 10^9$ J
1 tep	$= 1.08 \times 10^{10}$ cal $= 4.54 \times 10^{10}$ J
1 tec	$= 7 \times 10^9$ cal $= 2.93 \times 10^{10}$ J
1 m^3 gas	$= 3.94 \times 10^7$ J
1 t (TNT)	$= 10^9$ cal $= 4.184 \times 10^9$ J
1 Meu	$= 10^{18}$ J
1 kg material	$= 8.99 \times 10^{16}$ J

Energy symbols. J (joule, basic unit of the universal SI measuring system); kWh (kilowatt·hour); cal (calorie); erg (unit of physics); eV (electron volt); Btu (British thermal unit); quad, Q (multiple units of the Btu)) bbl (barrel of oil, corresponding to 159 liters or 42 U.S. gallons); tep or toe (ton of petroleum or oil equivalent); tec (ton of coal equivalent); m^3 gas (cubic meter of medium quality natural average gas at 1 atm); t (TNT) (ton of explosive, TNT equivalent); Meu (metric energy unit).

Prefixes. k (kilo, one thousand, 10^3); M (mega, million, 10^6); G (giga, billion, 10^9); T (tera, thousand billion, 10^{12}); E (exa, 10^{18}). For instance, in the case of watt·hour: Wh; kWh; MWh; GWh; TWh.

British and American units. Mass: 1 long ton = 1016 kg, 1 short ton = 907.2 kg, 1 pound = 0.454 kg. *Volume:* 1 U.S. gallon = 3.785 liters, 1 imp. gallon = 4.546 liters, 1 barrel = 159 liters. *Power:* 1 horsepower (hp) = 0.7457 kW, 1 metric horsepower (CV) = 0.7353 kW.

energy (consumed or generated) per unit of time. The main unit is the watt (symbol W), which is defined as the power equivalent to 1 joule per second. It is interesting, and in fact surprising, to learn how the concept of energy—today having such a fundamental and *Tortous* precise meaning in both physics and the practical world—came *path* along through a slow and tortuous path.

The early energy concepts were based essentially on Newton's famous "law of motion" as postulated about 300 years ago; although the concept of energy was first used by Kepler around 1620, even if with a still rather vague meaning. At that time it was nearly exclusively related to mechanical energy and thus included the kinetic energy of moving bodies (given by the simple expression $1/2\ m \times v^2$, i.e., one-half the product of the body's mass m times its velocity squared, v^2) and potential energy. The latter referred particularly to *A falling* the position of masses in the gravitational field (the apple falling *apple* from the tree) and also to other forms such as the potential energy of a compressed spring. The concept of mechanical work could be related directly to mechanical energy and thus used as a synonym for energy as defined by motion concepts.

With the foundation of thermodynamics in the nineteenth century, *Laws* the links between heat and mechanical work were established by the *and* laws of thermodynamics, thanks mainly to the work of Rankine, *order* Thomson, Young, Clausius, Boltzmann, and others.[1.6] The first law of thermodynamics establishes the equivalence of heat and mechanical work and postulates conservation of their sum in a closed system (Fig. 1.2). In addition, the kinetic theory describes heat as the result of the random motion of the mass' consituents (atoms, etc.; see also in Sec. 5.3).

The second law has to do with energy transformation efficiency and states that mechanical work can be fully transformed into heat ("thermal energy"), but on the contrary, heat can be transformed into mechanical work at the very best only with an efficiency expressed by the ratio of the temperature difference over the initial temperature, $(T_1 - T_2)/T_1$, in which T_1 and T_2 are the absolute temperatures* of the hottest (inlet) and coolest (outlet) parts of

---

* The absolute temperature is expressed in (degrees) Kelvin (symbol K) and is obtained by adding the number 273.16 to the degrees Celsius. The Kelvin temperature scale thus corresponds to the Centigrade scale shifted so that 0 K, the zero absolute temperature, is at minus 273.16°C.

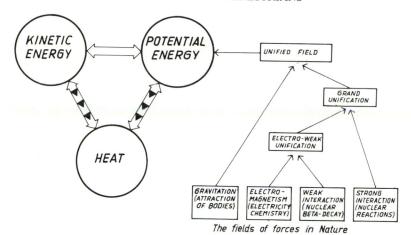

The fields of forces in Nature

Figure 1.2   The basic forms of energy and their transformations.

Kinetic energy is the energy of moving bodies. Potential energy is related to the concept of field of forces, because these enable an object to acquire energy when moving in them (e.g., the falling object within the gravitational field). Potential energy thus includes such forms of energy as electricity, gravitation, and compression of solid matter (e.g., a compressed spring). Modern physics wants to unify the four forces in nature into only one or two basic ones.

the transforming thermodynamic system. Example: The ideal, maximum efficiency of a steam turbine with an inlet steam temperature of 550°C and an outlet temperature of 50°C is (823.2 − 323.2)/823.2 = 60%; this means that in the turbine, at the very best, only 60% of the heat input is transformed into mechanical energy, the remaining 40% of the heat being discharged back into the environment. (Modern steam turbines achieve a net efficiency of about 45%; for an automobile engine it is 25%.)

In the second half of the last century the concept of potential energy was extended to include electrical energy, as it was estab- *Utility's* lished by Joule and others that an electrical current transports *bill* energy and can heat matter. In fact, the energy E in units of joule is expressed by the product $E = V \times I \times t$, where the potential V is measured in volts, the current I in amperes, and the time interval t, during which the current is flowing, in seconds.

In 1905 Einstein's theory of relativity, which expanded the earlier work of Voigt, Lorentz, Poincaré, and others, postulated the equiv- *A pound* alence of energy (E, in joules) and matter (m, in kilograms) through *of energy*

Figure 1.3   Energy, as seen by an advertisement at the end of last century.

the famous relation $E = mc^2$, where the coupling constant c, the speed of light, is equal to 300 million meters per second. Every increase of kinetic energy encompasses, according to relativity, an increase of the related object's mass: a fundamental law which became important and measurable with the large energy densities present within the atomic nucleus.

The concepts of indestructible and uncreatable energy have been further refined by modern physics. Energy is closely related to the field of forces because they contain (potential) energy. In fact, when a force causes an object to move, energy is being transferred from the force-related field into the object's kinetic energy. Present-day physics considers that in nature there are only three (possibly two, *Unified* or hopefully one) different kinds of force, namely, (1) the gravi- *dream* tational force, which manifests itself with large masses of celestial objects; (2) the electroweak forces, which include electromagnetic forces (responsible for electrical and chemical effects) and weak nuclear forces such as those that regulate the so-called beta decay of the nucleus; and (3) the strong nuclear forces which apply to the fundamental heavy particles in the nucleus and to their transfor- mations (Fig. 1.2).

In conclusion one can say that the concept of energy evolved only gradually within the general development of science and emerged in the last century over other thermodynamic quantities as being a significant and basic quantity (Fig. 1.3). Since then it has gained further importance in relation with electrodynamics, relativity and quantum theory. Nevertheless, even if energy is expressed with one unit only, the joule, it needs some specification as the second law of thermodynamics clearly indicates. Accordingly, one joule of thermal *A joule* energy is in physical and practical terms quite different from one *not always* joule of mechanical or electrostatic energy, as the latter two can be *a joule* fully transformed into the former but not vice versa.

For example, if the goal is to produce electricity through a rotating generator, then a unidirectional mechanical energy source (flowing water, wind, waves) provides a much more valuable energy to drive the generator than heat, as the latter is transformed with an efficiency of about 40% into electricity, whereas the former has efficiencies of typically 70% or more (Fig. 1.4). Unfortunately, man's major, primary energy sources (solar, chemical, and nuclear) basically provide heat.

The laws of thermodynamics determine and constrain the pro-

EFFICIENCY ELECTRICITY PRODUCTION

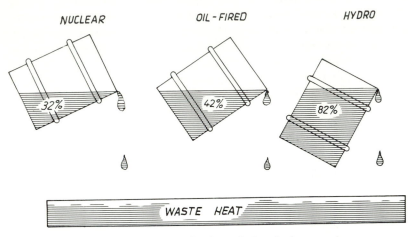

Figure 1.4   Efficiency in the production of electrical energy.

The efficiency in a thermal cycle depends on the operating temperature, as mentioned in the text; present nuclear power stations operate at lower temperatures than oil-fired ones and thus have lower efficiencies. Thus in such a power station only about one-third of the primary thermal energy is transformed into electrical energy; the rest is discharged into the environment as low-temperature waste heat.

Physics and economics cesses by which energy is transformed into useful work, and thus directed to the needs of man. In other words, these laws also govern the connections between physics and economics, and thus in a modern sense determine the social production, exchange, and consumption of energy, goods, and services.

Energy, food, and man

Human beings need food—a source of nutrition and energy—to grow, exist, and work. The average daily input of energy through food is 2 500 kcal, equivalent to 3 kWh or 10 million joules. The average power consumed by a man over 24 hours amounts, there-*Not even a horse* fore, to (3:24 =) 0.13 kW, which corresponds to approximately one-sixth of a horsepower (1 hp = 0.746 kW). Primitive man obtained his food by hunting, nomadic herding, and shifting agriculture.

Modern man gets it through more complex food systems (Fig. 1.5).

A study made some 10 years ago on the primitive community of the Tsembaga people in New Guinea found that the biomass output gives more than a 16 to 1 return on the human energy invested in a very primitive agriculture (by referring to Fig. 1.5, the energy output E divided by the human input B equals 16, as the external energy input C is zero). *Happy people*

Although this food system based on human labor alone requires the lowest energy inputs, it depends on the use of extensive areas of land per capita. With the growth in population, evolution has been toward more intensive agriculture and the employment of draft animal labor; therefore, the ratio becomes smaller, typically 5 or 3. In considering the energy balance of the food system as schematically

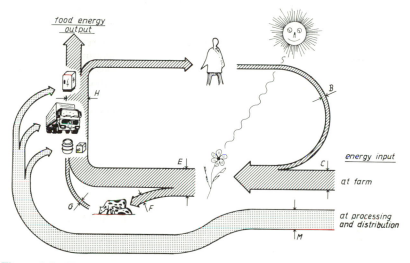

Figure 1.5   Food system managed by man.

Primitive man uses only his own energy input (in addition to the solar energy, direct or already transformed in the soil, not counted in this energy balance). Intense farming with high yield per unit of land surface requires large energy inputs (C) in the form of fuel, fertilizers, irrigation, and machinery. Processing and distribution of food as eaten by man requires additional energy. Typical energy ratios for intensive (American) agriculture and modern food systems are:[1.1] $C/E = 0.5$ for corn, $= 0.15$ for rice; $C/G = 1$ for milk, $= 3$ for eggs, $= 3$ for grass-fed beef, $= 15$ for feedlot beef; $M/C = 4$; $F/G = 15$. For corn in absolute value the annual input C can be 4 MJ/m^2, and the energy content of the output $E = 8$ MJ/m^2.

Intensive corn shown in Fig. 1.5, one actually neglects the crucial solar energy input upon which the entire food supply depends. With the advent of machinery, agriculture became ever more intense and required the use of more fertilization and larger energy inputs. For intensive corn or soybean agriculture the energy ratio between food output and energy input goes down to about 2.[1.1]

The farmer's cartel On this basis, the slightly less than 40 million tons of grain per year that the Soviet Union has imported on average since 1979 (about one-third from the United States alone) corresponds to about one-eighth of the 160 million tons of oil that the U.S.S.R., the world's second-largest exporter of oil, has exported annually since 1979. This enormous energy investment contributes to giving the United States a monopoly in grain, maize, and soya, which is nearly as extended as that which the oil-producing countries of the OPEC cartel have on oil.

Processed calories In carrying out an energy analysis of the food system, it is important to specify the composition of the food that provides the calories; for example, in order to obtain 1 kg of animal protein, 10 to 20 kg of vegetable protein are required, thereby reducing substantially the efficiency of the system. (In Fig. 1.5, G would amount to from one-tenth to one-twentieth of F.) Today, hardly any food is eaten as it comes from the field. Most of the food is processed, packaged, transported, distributed, stored, or frozen. All these steps eat up a large amount of energy, in the mean, as much as seven times the energy content of the food itself (in Fig. 1.5 this corresponds to the ratio of M over H).

Energy and survival The elements given here show the interrelation between energy and food, which in certain poor countries becomes the relation between energy and survival. As long as the cost of energy is increasing, it would seem that intense high-yield agriculture, as practiced in Western Europe and in the United States, is not necessarily a good solution for the poor of this world; for example, to fill the entire world with a U.S. type food system, about three-quarters of the present, annual world energy consumption would be required.[1.1]

1.3   FROM SOURCE TO CONSUMPTION

Generally speaking, one can say that man requires energy, food, and materials for his physical existence on earth:

- Energy, for producing heat and doing work
- Food, for his nutrition (with the energy aspects discussed in the previous section)
- Materials, for the production of food, for private and public construction works, and for transporting, transforming, and recuperating energy (again strongly energy related)

The energy consumed by man for these requirements is derived from some primary sources[1.8] which at present include principally fossil energy (coal, oil, and gas), and some minor contribution to

The man-energy equation

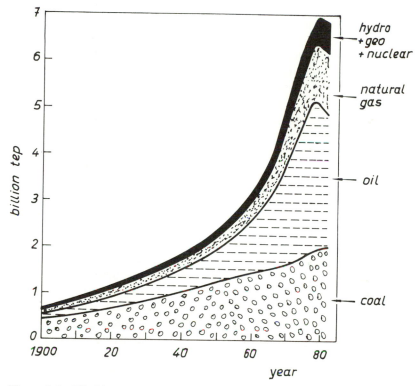

Figure 1.6   World energy consumption from primary sources in the twentieth century, expressed in tons equivalent petroleum per annum.[1.17]

The 1982 (1983) consumption of 6835 (6925) million tep could also be expressed by an equivalent 140 million barrels of oil equivalent daily, or 5 liters per capita per day. Maximum total consumption has been reached in 1979 with 6935 million tep. In 1982 (1983) the relative shares for the four categories were (from top to bottom): 663 (705), 1312 (1329), 2819 (2794), 2041 (2097) million tep.

electricity production by nuclear energy and hydroenergy (Figs. 1.6 and 1.7). The use of direct solar energy is still irrelevant and that of wood for burning[1.12] is important only in some third world regions, but not on a global basis.

It is significant that oil and natural gas, both of which are destined to disappear as a major source of energy in 50 to 100 year's time (depending on consumption levels and reserves, see Fig. 4.2), today (1983) cover 60% of the world's energy requirements. Coal and lignite, which as recently as 1953 still provided half this requirement, today account for 30%. Therefore, fossil sources together *A changing* cover 90% of world consumption at present. All the other sources *world* (nuclear, hydroelectric, geothermal, solar) which when oil and gas will decline should cover over 70% of the total requirement, today contribute 10%. These figures give some idea of the huge problems involved in the renewal of technical structures which will have to be faced over the next 50 years and beyond.

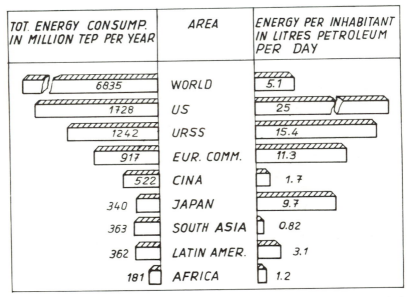

TOT. ENERGY CONSUMP. IN MILLION TEP PER YEAR	AREA	ENERGY PER INHABITANT IN LITRES PETROLEUM PER DAY
6835	WORLD	5.1
1728	US	25
1242	URSS	15.4
917	EUR. COMM.	11.3
522	CINA	1.7
340	JAPAN	9.7
363	SOUTH ASIA	0.82
362	LATIN AMER.	3.1
181	AFRICA	1.2

Figure 1.7 Primary energy consumption in some geographic areas of the world at mid 1982.[1.17]

As usual, primary electrical energy (from hydroelectrical or nuclear power stations) has been converted into tep using the calorific equivalent of the fuel which would have been necessary in a conventional power station to produce the electricity, i.e., roughly 210 grams of equivalent petroleum for 1 kWh electrical energy.

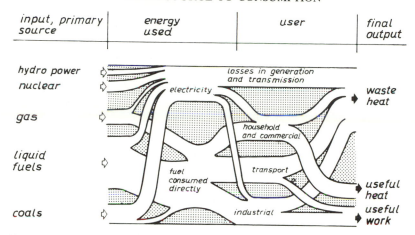

Figure 1.8   Typical energy-flow diagram for a closed system or country, where input includes various local sources and import/export balance.

On a world basis,[1.17] total energy consumption at mid-1983 was 6 925 million tep, where the primary sources contribute (from hydro + nuclear to coals): 705, 1329, 2794, 2097 million tep.

   The energy-flow diagram in Fig. 1.8 shows that the primary energy is mostly used in the form of convenient fuels and of electricity. Because of the limited efficiency of the energy transformation and consumption processes, about two-thirds of this goes directly into waste heat and only the remaining one-third produces useful heat *A lukewarm* and work. At the end, as a consequence of the laws of thermodyn- *end* amics, even this third will degenerate into low-temperature heat, which seems to be the end stage of any energy form in the physics world we know today.

   The quantity and quality of consumption varies markedly from one country to another, even among the western industrialized nations. The United States is the biggest consumer, at present (1984) accounting for about 25% of energy consumption, even though *Products* it has only 5% of the world's population. As can be seen from *and* Table 1.2, the energy consumed per capita varies substantially. *energy* Equally significant is the variation of the average consumption based on the gross domestic product. This GDP is a useful index of the level of economic activity of any one country, since it expresses the market value of the final products and services provided by the national economy.

Table 1.2 Gross domestic product and energy consumption per year (mid-1982)

Country	Population[5,18] (millions)	Gross Domestic Product (GDP)[1,16] ($/inhab.)	Total Energy Consumption[1,17] (tep/inhab.)	Electricity Production (MWh/inhab.)	Energy Consumption per Unit GDP (kg eq. petrol/$)
United States	232	13 050	7.4	10.1	0.56
European Community	271	8 660	3.3	4.4	0.38
Australia	15.2	10 450	5.1	6.3	0.48
Canada	24.6	11 760	8.4	14.9	0.71
Japan	118	8 850	2.8	5.1	0.31
Germany (F.R.G.)	61.6	10 660	4.0	5.9	0.37
France	54.2	9 940	3.3	4.7	0.33
Italy	56.6	6 140	2.4	3.2	0.39
United Kingdom	55.8	8 400	3.4	5.0	0.40
Belgium	9.9	8 380	4.6	5.4	0.54
Denmark	5.1	10 960	3.5	4.9	0.31
Netherlands	14.3	9 660	4.4	4.5	0.45
Greece	9.8	3 840	1.7	2.2	0.44
Austria	7.6	8 780	3.2	5.5	0.36
Finland	4.8	10 060	4.4	8.0	0.43
Norway	4.1	13 760	7.7	20.4	0.55
Portugal	10.1	2 280	1.1	1.4	0.48
Spain	37.9	4 760	1.9	2.9	0.39
Sweden	8.3	11 880	4.5	11.6	0.37
Switzerland	6.5	14 740	3.9	7.4	0.26
Turkey	46.3	1 130	0.6	0.5	0.53

Another parameter that is often used to analyze and forecast energy consumption is the so-called energy coefficient, or elasticity, which is defined as the growth rate of primary energy consumption divided by the growth rate of GDP. An energy coefficient of 1 (which was the mean value in the western world for the years 1960−1973) means, for example, that a 4% annual increase in GDP requires an increase of 4% in energy consumption. The coefficient tends toward higher values when industrial processes are wasteful, fuels are used less efficiently, or domestic consumption is excessive. In the next decades mean values of 0.6 or, hopefully even less, would seem desirable. *Desirable elasticity*

Most of the data given so far include only the explicit sources and consumption of energy (oil, coal, nuclear energy). However, as mentioned previously, the extraction, cultivation, or processing of the most diverse products require an energy input alongside the raw material. It is the goal of an important new field of study—known as energy analysis—to evaluate qualitatively and quantitatively the various forms of primary energies required for the production of commodities.[1.19] Such an analysis will be influenced by other related areas of study: overall energy policy, resource studies, environmental studies, political aspects, and the economics (price structure) of energy and raw materials. The analysis includes the direct use of energy in the production and also the energy required to produce and move materials and machines used in the process. Besides the direct result, the gross energy requirement, such an analysis could provide new insights and perspectives in the production process and its interrelated aspects (environmental, economical, etc.). *Energy analysis*

The gross energy requirement for some typical products is given in Table 1.3. For example, the table shows that importing 1 kg of meat is, from the energy point of view, equivalent on average to importing 1.3 kg of gasoline. A calculation along these lines would raise Switzerland's energy consumption above the level shown in the Table 1.2, since this country imports many energy-intensive products. Energy requirements for recycled metals are lower than those for primary production indicated in Table 1.3:[5.25] the production of iron from scrap requires about one-third the energy, for copper it is one-tenth, and for aluminium less than one-fifteenth. *Meat and gasoline*

Electricity is a particularly useful form of energy. An average 25% of the primary energy in industrial countries is used to generate electricity (Table 1.2). However, the limited efficiency of a con-

Table 1.3   Gross energy requirement (for mining, cultivation, refining, production, etc.).

	Million joules per kilogram (MJ/kg)
Metals[5.25]	
Iron (ore grade 60%)	30
Stainless steel 304[1.15]	110
Aluminum (25%)	270
Copper (0.6%)	110
Lead (4%)	30
Plastics[1.18]	
PVC (incl. feedstock: 23 MJ/kg)	70–90
Polystyrene (incl. feedstock: 46 MJ/kg)	100–140
Polyethylene (incl. feedstock: 46 MJ/kg)	80–120
Various materials[1.18]	
Glass	13–22
Rubber, natural	6
Rubber, synthetic	140
Cement	5–7
Soil, rocks, salt	0.5
Inorganic chemical products, paper, etc.	20
Organic chemical products, fertilizers, etc.	60
Food[1.1]	
Vegetable food stuffs	4–6
Animal protein (meat, fist, milk)	40–100
Calorific energy (for comparison)	
Gasoline	46
Water: Heat from 0 to 100°C	0.42
Melting (0°C)	0.34
Evaporation (100°C)	2.3

Electric connection ventional power station (Fig. 1.4) means that the electrical energy produced and used represents, in fact, only about 10% of the overall energy consumption. The share of electricity in the overall energy picture is expected to increase markedly in the decades to come (in the United States it has about doubled from 1968 to 1983). The total net installed capacity in the United States and in the European Community was (in 1982) 670 and 330 gigawatts (electric), and the electricity produced thereby amounted to 2 420 and 1 270 billion kWh (i.e., 2.42×10^{15} and 1.27×10^{15} Wh). Electricity is derived from a variety of primary energy sources. It is hoped to boost the present 10% derived from nuclear energy to about 30% or more by the end of the century. The global yearly production of hydro energy (i.e., by water driven turbines) amounted in 1984 to about 1 700 billion kWh; an additional 550 is under construction, and about 1 100 are potentially possible. The theoretical hydro energy of our planet of about 16 000 billion kWh per year would thus at the end be exploited by roughly 20%.

Transporting energy from the source to the consumer is one of the major problems in the energy economy as enormous quantities in weight and volume must be shifted often over thousands of *Moving* kilometers. All the major means of transportation are involved in *around* this task: marine, inland waterways, railways, and roads. Electricity is transported by high voltage lines (see Sec. 5.2).

Pipelines are the most important transportation system for liquid *A web of* and gaseous fuels. Particularly in the United States and in Europe, *oil* there is an extended web of oil pipelines. In Western Europe about 650 million cubic meters of oil and its derivatives were transported in 1979 through pipelines with an overall length of over 19 000 km. A recent, much discussed enterprise is the Trans Alaska pipeline built in a very difficult and environmentally delicate permafrost region.

The transport of natural gas in a gaseous form through pipelines is even more interesting, as the other valid alternative is to liquefy *Bubbling* the gas and transport it in containers, a process that is more energy- *pipes* consuming and poses practical safety problems. An impressive system of gasducts is in operation or under construction all over the world, particularly in the United States. The Trans-Mediterranean pipeline transports gas over 2 400 km from the Algerian Sahara under the Mediterranean Sea to northern Italy; from there another pipeline crosses the Alps to the Netherlands and Germany, where the pipelines from the gas fields of the North Sea also arrive.

In 1984 the 5 500-km-long gasduct from the gas fields of the Jamal peninsula in the northern Siberian permafrost regions to Vienna *From* came into operation. Its realization cost about $15 billion, and it will *Siberia to* deliver in a few years' time over 40 billion cubic meters of methane *the* gas to Western Europe each year over the existing European gas- *Eurokitchen* duct system. Discussions on this undertaking lasting over 2 years have demonstrated to the general public once again the kind of political and economic problems that such large energy projects entail. A detailed project exists for a gasduct, nearly 8 000 km long, from Alaska across Canada into the midwest of the United States. It is estimated to cost over $40 billion, of which about $30 billion are foreseen for the difficult 1 200-km-long crossing of Alaska.

Coal drops In addition to liquid or gaseous fuels, solid coal (and often metal ores, as well) is transported in pipelines.[1.5] The coal is crushed into small particles, mixed with water and thus formed into a slurry, which is pumped through the lines. The largest pipeline of this type in operation transports 5.5 million tons of coal per year from a coal mine in Arizona through 450 km to the Southern California Edison Company's Mojave power station in Nevada. Much longer coal-slurry pipelines are scheduled to go into operation by the end of this decade in the United States.

As we discussed in the previous section, a meaningful energy analysis also requires the evaluation of the intermediate transformation efficiencies and a detailed determination of the quality of energy consumption. Efficiency in the use of energy can be illustrated by the energy requirement of transportation.[1.4] The efficiency varies widely, for example, a bus carries a passenger with typically one third the energy as a private car; [1.3] but this is not the whole story—comfort and speed are valuable assets for public acceptance. Never-

Lucky traveler theless, as the movement of people and freight accounts for about 20% of energy consumption in the developed world, fuel economy of transport is an important argument in energy conservation. It is obvious that conservation can be obtained by shifting from private to public mass transportation, and by increasing the efficiency of private cars from typically 10 km liter of fuel to 20 or even 25 km liter (as anticipated for the 1990s) in realistic highway driving conditions (experimental one-man vehicles can travel in ideal conditions more than 600 km with 1 liter of fuel).

1.4 ENERGY AND RISKS

Hazards in the real world

While energy is, as we have already seen, the key to improvement in man's standard of living, its increasingly widespread use nevertheless entails some risks to man himself and the environment in which he lives. The debate in this context has hitherto been centered almost exclusively on the risks and drawbacks of nuclear energy (these will be examined in Chap. 3). Nevertheless, the quest for a *An* strategic solution to the energy problem must include an objective, *extended* comparative assessment of the merits and the hazards associated *debate* with the exploitation of all the major energy sources considered today.[1.11]

On the basis of arguments that will be discussed in the following chapters, it is likely—although opinions are far from unanimous— that the solution to the energy needs of the medium-term future lies in th   following four-point strategy:

1. A significant reduction in the consumption of energy in the industrialized nations, through conservation and/or more efficient transformation processes
2. Further application of nuclear energy for electricity production *Opinions*
3. Increased production and consumption of coal and natural gas *and strategy*
4. Development and deployment of alternative energy sources, i.e., solar energy and minor sources such as geothermal, wind, and tide energy

This strategy will require further technological development in support of the large-scale production processes that transform energy from one form into another (e.g., electricity from nuclear energy); however, new small-scale technologies for the efficient and flexible production and consumption of energy at the relatively small consumer level will also become more important.

Risk assessments of any of these present-day or future energy-related technologies are difficult and often questionable since, un- *Weighting* like energy's economic costs and benefits, they do not lend them- *damages* selves to easy characterization.[1.14] In addition, the concepts of risk analysis (as hazard probability, or expected harm, this being the sum of all possible damages weighted by their probabilities) have to be checked against the risk perception of the public, a delicate and complex, if not controversial, issue.

Most people just have difficulty in appreciating objectively the issue fo risk. Faced with the complexities of the real world, people (and the mass media) tend to concentrate on only one or a few simple issues, thereby often ignoring the vital issues imbedded in complex problems. This, of course, raises the challenge of an adequate education and divulgation, since even if public opinion remains subjective or emotive, it nevertheless will ultimately shape—

A democratic challenge in a democratic system—society's preferences through the political process.

One first step to present the issue of risk is to distinguish between occupational and environmental hazards. The latter are taken in the broadest sense to include sociopolitical as well as biological and geopolitical aspects. A comparison between various energy sources requires the analysis to encompass the whole energy supply and production chain of each single source, such as the complete fuel cycle (including transport and reclaiming or reprocessing) and the construction and operation of the power plant.

Occupational hazards, i.e., hazards experienced by workers operating within the energy chain, are understandably the easier ones to analyze. As a consequence, many more studies have been done in this field. But even here there are many questions open to *Worker's hazard* debate. For example, what should be the relevant hazard parameter of comparison: hazard per worker employed or hazard per unit energy produced? Coal would compare unfavorably with all other major sources if the latter were the parameter of comparison, because the mining and transport of the large coal masses inevitably produce a relatively large number of hazards. On the other hand, coal is a labor-intensive business, and thus the hazards per employed worker may not result in such a bad comparison. In a period of high unemployment, one would thus probably take this as the proper parameter.

Environmental hazards are more difficult than occupational hazards to assess on a comparative basis. In addition, indirect hazards that are difficult to predict are often perceived to be more *Environmental hazards* important than direct ones. For example, with respect to fossil fuel energy, one is becoming increasingly aware that climatic changes due to increased carbon dioxide emission and the acid rain effects on terrestrial ecosystems could pose greater threats to human well-being than the easier to detect, direct hazards of accidents.

Comparative risk assessment is too complex a subject to be considered further in this section. In view of the issue of fossil energy (in particular, increased coal usage) versus nuclear energy we want to mention, nevertheless, some possible long-range effects of the burning of fossil fuels. The risks related to nuclear energy will be discussed in Sec. 3.3.

Acid rain in the greenhouse

One of the significant sources of energy for the future is coal, since the available reserves are enough to satisfy our needs for more than a millenium (see Chap. 4). In addition to the obvious ecological hazards inherent in its mining process, particular mention must be made of the problems of carbon dioxide and acid-rain precipitation related to the combustion of any fossil fuels. These are environmental hazards which still require in-depth investigation and which could ultimately prove extremely serious.

Carbon dioxide (CO_2) is vitally important to the biosphere, inasmuch as it is an essential ingredient of photosynthesis and, therefore, supports life on our planet both directly and indirectly. Today it accounts for 0.03% of the atmosphere. Will the millions of tons of CO_2 given off by the ever-increasing combustion of fossil fuels and released into the atmosphere ultimately increase this amount, or will the biosphere's CO_2 cycle and the absorption of CO_2 by the oceans succeed in keeping the rise in check? Some data will help to qualify the problem. *Vital and toxic*

The waters of the oceans contain about 40 000 billion (4×10^{13}) tons of dissolved carbon, the biosphere 2 000, and the atmosphere 500 billion tons. Annually, the exchange between these reservoirs amounts to 200 billion tons, whereas the immission into the atmosphere caused by man through combustion is about 5 billion tons. The increase of the carbon dioxide content in the atmosphere is thus mainly an effect of how the exchange works between the reservoirs. However, the fact remains that the percentage is at present increasing. But why is this dangerous? *Useful reservoirs*

Carbon dioxide (as well as water vapor) allows the sun's rays to penetrate the atmosphere, but prevents part of the reflected infrared radiation from leaving it. The result of this property is to heat the surface of the earth as if this were a greenhouse (see by analogy the solar collector of Fig. 4.10). Extrapolating on the basis of the present rate of increase, the carbon dioxide content of the atmosphere could even double at the end of the next century (between 2 050 and 2 100), resulting in a rise in the global temperature of the earth's surface of 1.5 to 3°C according to recent estimates[1.13] (note that during Ice Age the earth's surface was globally only about 2.5°C cooler than it is today). Increased use of coal could thus gradually change the climatic and meteorological conditions quite dramatically. *No flowers in the greenhouse*

There could be significant changes in agriculture, and, in the extreme case of a long enough warming, the West Antartic ice sheet could collapse into the sea with a consequent 5- to 6-m rise in sea level.

In conclusion, one can say that the quantitative aspects of the carbon dioxide problem due to increased fossil fuel burning is still open to debate. However, the potential danger it spells imposes considerable caution in handling this problem. An extreme example *A hot* of the greenhouse effect is provided by the atmospheric conditions *planet* on the planet Venus. Its prohibitively high surface temperature of about 500°C is mainly the result of the greenhouse effect due to the combined action of carbon dioxide and water vapor which constitute the atmosphere of that planet.

Another concern related to an increase of fossil fuel burning both for heating and transportation is the so-called acid-rain problem. *Walking* The combustion of fossil fuels annually releases over 100 million *in the* tons of sulfur in the form of sulfur dioxide and 6 million tons of *acid* nitrogen in the form of various oxides into the atmosphere of the *rain* northern hemisphere alone. A part of these products appears as dry deposition near the source, but the rest can be further oxidized into sulfuric and nitric acids, which then precipitate as rain, causing a long-distance pollution thousands of kilometers away.

There is as yet no comprehensive understanding of the effects of acid deposition on terrestral vegetation and the biosphere in general, although in some instances researchers have been blaming it for killing fish, threatening vegetation, and damaging metal and stone in monumental structures. It is a fact that the precious forests of Western Europe have shown alarming signs of illness in recent years. At mid-1984 more than one-third of West Germany's forests, for example, were affected to some degree; the evergreens have been the most susceptible, with three-quarters of the country's firs being affected. Recent research, while confirming that this damage is directly related to the increasing atmospheric pollution, has also shown that the problem is more complex than originally thought.

It should also be mentioned that in addition to man-made emissions *Men and* of sulfuric and nitric acids there is also natural emission of these *volcanoes* products. For example, emission into the atmosphere of sulfuric acid from volcanoes and through bacterial reduction of continental shelf sediments is estimated to be in the range of 40 to 110 million tons a year, i.e., about comparable to the man-made emission.

Finally, it may be instructive to mention a problem that relates in a way the safety of fossil-fuel-burning power stations to those using nuclear energy: the emission of radioactive elements through coal burning. In fact, coal contains traces of uranium, thorium, and their decay products which in combustion are either deposited in the *Radioactive* ashes or emitted into the atmosphere. The amounts change from *coal* one geological deposit to another.

In Germany recent measurements have shown that the radio-active emission of a coal-burning power station is, on average and under normal conditions, seven times as high as that of an equivalent nuclear power station, still remaining well within accepted safe emission levels. (It should be noted, however, that for nuclear energy, the prevalent radioactive emission comes from the full nuclear fuel cycle outside the power plant; see in Chap. 3).

CHAPTER 2

Energy in Mankind's Evolution

Man's progress and the development and spread of civilization have in the past been in direct relation to the quantitative and qualitative control of new and more sophisticated sources of energy. Will the same be true with novel forms of energy, in particular with nuclear energy? Before we address the evolutionary aspects concerning energy in the last decades and centuries, we look back to the origins of terrestrial energy sources. We will discover a spectacular story, the beginning of which dates back to the very first events following the creation of the universe.

2.1 ENERGY IN THE UNIVERSE

Big Bang and fusion

Investigation into the origin of the elements that provide the practically unlimited terrestrial energy resources leads us to discuss the evolution of the universe and the origin of the solar system, exciting topics that are still open to scientific debate. There is at present a *Explosive* remarkable convergence of experimental evidences and opinions, *birth* which relate the origin of the universe at somewhat more than 15 billion years ago to a primordial explosion, the so-called Big Bang.

According to this theoretical picture, all elementary particles could be created, in reversible transformations among them and into radiation, out of the enormous amount of energy concentrated in the Big Bang (estimated at 10^{68} J). However, as the universe began to expand rapidly and thus to cool down, the reversibility between particles and radiation gradually ceased, and part of the energy remained condensed into particles, according to the concepts of the theory of relativity. This occurred first for the heavier particles, such as the protons and neutrons, and then for the lighter ones, such as the electrons (see Fig. 2.10).

In this way, after a fraction of a second from the origin, as the temperature fell from 10^{13} K toward 10^{11} K,* energy condensed into primordial protons and neutrons. These could undergo many re- *Primordial* versible reactions among themselves and also with other particles, *protons* but could not, at that stage, be transformed back into radiation. Later, when the universe was about 10 min old and hence at temperatures of 10^9 to 10^8 K, all that remained possible was a basic nucleosynthesis whereby a free proton sticks to a neutron to form a deuteron (see reaction 1 in Ref. 2.1; also Fig. 2.2).

The deuterons thus produced could have been consumed nearly as fast as they were created by a variety of subsequent nuclear fusion reactions, finally resulting in the formation of helium nuclei (see reactions 2 to 6 in Ref. 2.1).

The energy-producing synthesis of helium is a basic process in the energy balance of the universe; even if it was mainly achieved in the eventful first 10 min after the Big Bang, this reaction is still going on in the stars, as we shall discuss later. Helium is an abundant element in the universe—about 1 for every 10 hydrogen atoms or protons— because, being a very tightly bound and stable nucleus, once formed it hardly undergoes any further transformation.

The rapid cooling and density reduction caused by the continuing expansion of the universe brought the binary deuteron-consuming reactions to a practical stop approximately 20 min after the Big Bang. As a consequence, a small fraction—one deuteron for every 70 000 protons—has been preserved to the present time. On earth, *Deuterons* the fraction is higher by one order of magnitude than in the universe: about one deuterium atom (containing the deuteron as its nucleus) for every 6 700 hydrogen atoms! This relatively high abundance is put into relation with what is considered today the most likely formation process of the earth, the gravitational self-attraction of gas and dust, including solid particles, that resulted from a previous explosion of a celestial object (a supernova). As these swept through space, the heavier deuterium of the interstellar gas would have shown a greater adhesion than hydrogen, thus resulting in the concentration effect on the matter that finally converged to form the earth.[2.7]

In anticipating the description in Chap. 5, it should be pointed out that deuterium represents a practically inexhaustible terrestrial

---

*K stands for the Kelvin temperature scale (see footnote in Sec. 1.2).

energy source, to be exploited through the controlled nuclear fusion process. The nucleus of deuterium is the basic and lightest nuclear fuel block. It is relatively easy to have it react, and thereby to *A miracle* produce energy, but it is very difficult to create; and this is the reason why it is so rare. By a miraculous physical effect, a tiny fraction of the energy available at the creation of the universe has thus been preserved in this element (and even concentrated on earth) up to the present time for use by mankind.

As the decreasing temperature of the expanding universe fell below 5 000 K some 1 million years from creation, an important new fact emerged. The nuclei, having positive electrical charge and being less violently shaken in their thermal movement by the de- *From* creasing temperature, were able to catch the free, negatively charged *atoms to* electrons, and thus to form neutral atoms. The proton with the *paradise* electron were able to form a hydrogen atom, the deuteron a deu- terium atom, and so forth (see Fig. 2.10). There was at that stage an immediate important consequence: As neutral atoms do not interact as efficiently with radiation as the free electrically charged nuclei and electrons, the weak gravitational attraction was able to prevail even in diluted gases. This made possible the formation of galaxies, the first substantial condensation process, from which a series of new events emerged: the formation of heavier elements, the for- mation of the sun with its planets, and finally the creation of life on earth.

Supernovae and fission

Elements heavier than helium, with the possible exception of some lithium and beryllium, were synthesized not in the Big Bang but much later inside certain stars. This refers in particular to uranium and thorium, which were created in completely different circum- stances to those of deuterium.

Many million years after the Big Bang, matter (mostly in gaseous form) of the universe condensed by gravitational attraction into *Galaxies* galaxies of various sizes and shapes (Fig. 2.1). Galaxies are a fundamental element in the organization of the universe; they also represent an extremely interesting and complex accumulation of energy. A galaxy contains enormous amounts of gravitational energy, random kinetic energy, and (in many, but not all cases)

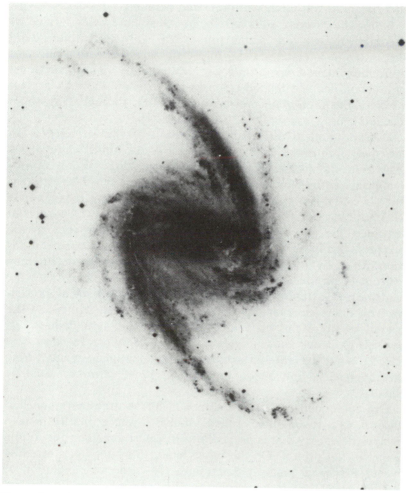

Figure 2.1   In the arms of spiral galaxies (this galaxy NGC 1365 is in the constellation Formax) young stars are forming continually by gravitational condensation of interstellar gas and dust.

The nuclear synthesis takes place inside these stars; it transforms primordial protons into helium and then, through various reaction steps, into other elements up to iron. (Courtesy: European Southern Observatory.)

rotational energy. In addition, in the center of some galaxies (as in our Milky Way), there seems to be a huge concentration of mass and energy in still unknown forms and amounts.

Within the galaxies, stars have been and still are being formed by gravitational attraction of the primordial hydrogen (protons) and

Star dust also traces of interstellar dust. Inside the stars (as within our sun) sufficiently high temperatures are produced by this gravitational contraction and the related compression to allow the nuclear synthesis of the primordial protons into helium, through the proton reaction chain (reactions 7 to 10 in Ref. 2.1).

The first step in this chain, the fusion of two protons, is an extremely rare event, as it proceeds about 10^{18} times more slowly than a strong nuclear reaction (for example, the fusion of two deuterons) at the same density and temperature. Why? The nuclear

Thermonuclear hangup force between two protons fails by only a few percent in magnitude to produce a bound state which would be helium 2. As a consequence, the only other possible reaction channel is that, as the two protons approach each other and start to interact, one proton is transformed by a weak interaction process into a neutron, thereby emitting a positive electron (and a nearly mass-less particle, the neutrino); this transformation process then allows a strong proton-neutron interaction leading to the formation of a deuteron. The overall result of this encounter of two protons is a deuteron, a positive electron, and energy (also a neutrino; see reaction 7 in Ref. 2.1), but the required sequence of the mentioned events during the infinitesimal interaction time makes it an extremely unlikely event.

This consequence of weak-interaction forces—the thermonuclear hangup, as it is sometimes called—makes it impossible for man to exploit the energy that is packed into the most abundant fuel in the universe, the proton. On the other hand, this hang-up is essential to our existence, since it determines the "smooth" evolution of the universe and its composition. Here again, a small physical effect (the just missing magnitude in the proton attraction to form

Figure 2.2 Evolution of the universe with the periods when energy was packed into protons, deuterium, and lithium (nuclear fusion), into uranium and thorium (nuclear fission), and into fossil fuels. The condensed masses experience different phases of expansion (driven by large energies set free) and of condensation (driven by gravitational attraction).

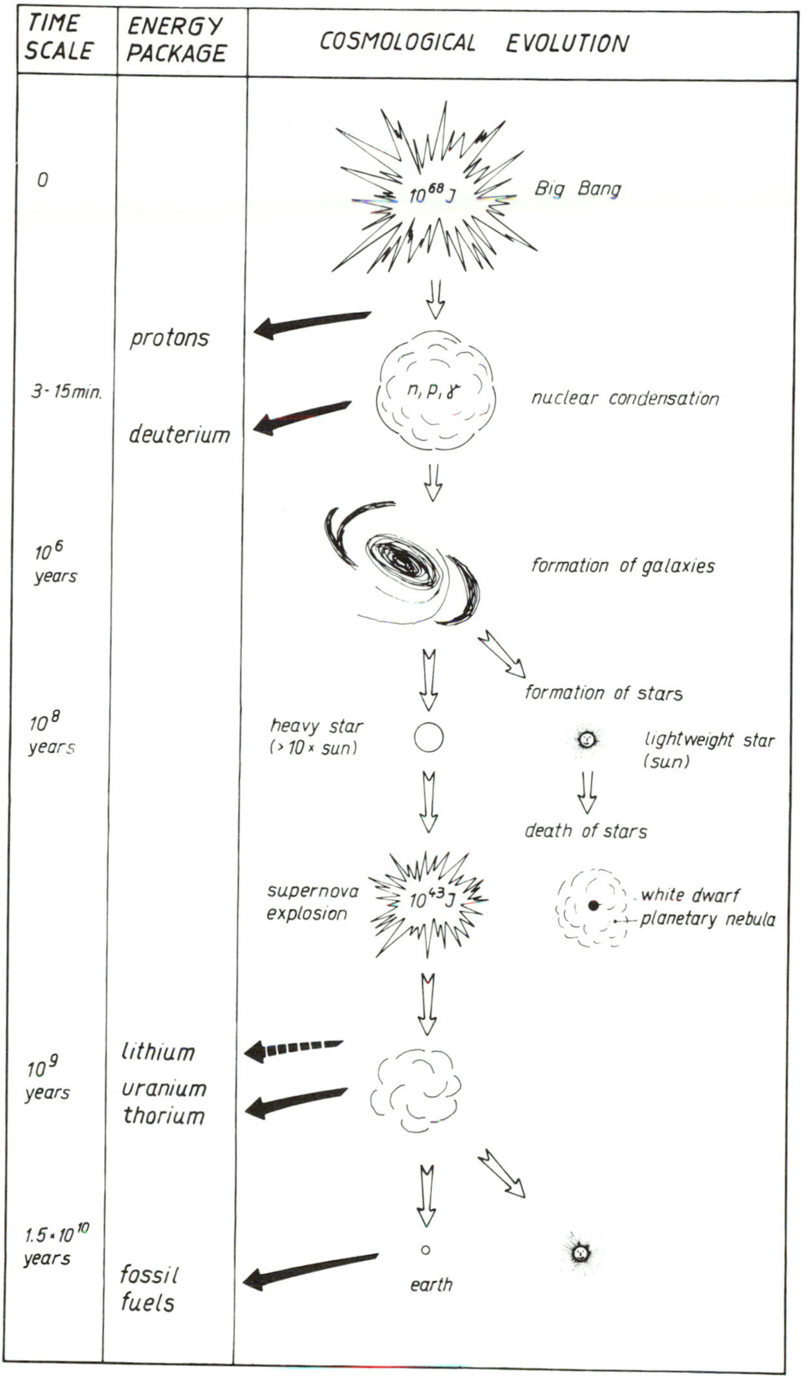

TIME SCALE	ENERGY PACKAGE	COSMOLOGICAL EVOLUTION

TIME SCALE

0

3 - 15 min.

10^6 years

10^8 years

10^9 years

$1.5 \cdot 10^{10}$ years

ENERGY PACKAGE

protons

deuterium

lithium

uranium
thorium

fossil
fuels

COSMOLOGICAL EVOLUTION

10^{68} J   Big Bang

n, p, γ   nuclear condensation

formation of galaxies

formation of stars

heavy star
(> 10 × sun)

lightweight star
(sun)

death of stars

supernova
explosion 10^{43} J

white dwarf
planetary nebula

earth

helium 2) has the most important, providential consequence of allowing the existence of the world as we know it today.

There are basically two types of star evolution (Fig. 2.2). Single stars having approximately the sun's mass or less (the most frequent stars) evolve smoothly through various cycles of their life, as their *Dead of* central supplies of protons are fused into helium (and, marginally, *a star* some slightly heavier elements). Finally, they shed their outer layers of gas, and fade slowly and dimly away as white dwarfs. In the case of our sun, the whole process lasts about 10 billion years, half of which is still ahead of us.

A much more exciting and rapid evolution is experienced by stars having masses of about five or more solar masses; they evolve toward a stage called supernova. The hotter and denser core obtained as a consequence of larger gravitational compression allows the critical step toward the synthesis of heavier elements: the nuclear reaction of helium. This requires the tricky, nearly simultaneous *Tricky* collision of three helium nuclei to combine into a carbon nucleus *collision* (reactions 13 to 16 in Ref. 2.1). From there on the synthesis proceeds in steps until iron is finally produced in the last nuclear fusion reaction that delivers reaction energy. Therefore, with the production of iron, the star has, so to say, run out of fuel.

At the end of the short life of these massive stars (100 million years or less), the star's interior consists of onionlike layers where different nuclear syntheses have subsequently occurred and are still taking place: hydrogen is still fusing into helium in the outer layer; but more toward the center helium reacts into carbon and oxygen, which in turn react into magnesium and silicon; and, in the core, into iron. This inner portion can become unstable by violent shrink- *Supernova* ing, as a consequence of gravitational forces, into a core of 100 000 times less its previous size, the so-called neutron star (or black hole for the most massive stars). This collapse releases in about 1 s more (gravitational) energy than all the energy our sun will put out in its whole life. In the remaining outer mass of the supernova, the process actually results in a violent explosion in which still heavier nuclei (by subsequent captures of neutrons) can be synthesized and then blown out into space. In our galaxy just five such events are known to have been observed in the past 1 000 years, but about 10 per year are now sighted in distant galaxies.

Supernovae, with their dramatic energy transformation processes, represent a determining element in the evolution of the universe. In

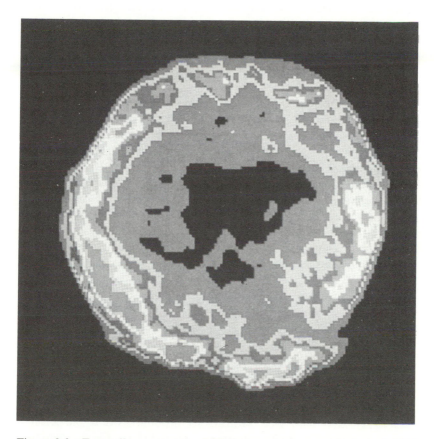

Figure 2.3 Expanding remnant of Tycho's supernova explosion of 1572 observed through a very large array of microwave receivers tuned at the wavelength of 22 cm.

The spherically expanding shell consists of material ejected by the explosion mixed with interstellar gas. Elements heavier than iron and up to uranium can be produced by subsequent multiple neutron captures in the supernovae explosion and are then blown out into the interstellar gas; later this mass including some heavy elements may eventually condensate to form new celestial objects. (Courtesy: Very Large Array Interferometer Facility, Soccorro, New Mexico.)

particular, protons and helium can be synthesized into the full range
of heavier elements. Probably more than one successive process of
this kind is required to produce the heaviest elements up to uranium.
The supernova explosion also distributes these elements into the
Uranium galactic space, where they are eventually included in subsequent
and generations of stars and their planets (Fig. 2.3). The energy that can
thorium be liberated today by nuclear fission on earth was packed into
uranium and thorium in just a few seconds of a supernova explosion,
more than 5 billion years ago!

A few words must be devoted to the generation of lithium which
has a galactic abundance of one for every billion hydrogen atoms (or
protons), and on earth is present with the isotopes lithium 6 in the
relative abundance of 7.4% and lithium 7 (92.6%). The two isotopes
of this element (see isotope definition, Fig. 2.10), together with
deuterium, represent the basic fuel of controlled thermonuclear
fusion, at least insofar as fusion will operate on the relatively easy
deuterium-tritium cycle (see Chap. 5). The origin of lithium is not
clear. There are various models describing its formation but no
single model prevails. It seems accepted that lithium 6 (as well as
Lithium beryllium 9, boron 10 and 11) is generated by the collision of
"cosmic-ray" protons with heavy nuclei in interstellar matter.[2.2] In
addition to such a creation mechanism, lithium 7 could also result
from the fusion reaction of helium 3 and helium 4 nuclei to give
beryllium 7, which by β emission (see note in Table 3.3) subsequently
transforms into the lithium isotope. This helium burning takes place
in the interior of stars, and, in a limited way, may also have taken
place in the Big Bang. However, since lithium 7 reacts nearly as fast
as it is created with free protons to produce two α particles, a
transient expanding situation must be postulated (as for the deuteron
production) whereby some lithium 7 nuclei survive because they are
simply not able to meet the escaping proton neighbors; such a
situation can be found during the explosive phases in the lifetime of
some medium-weight stars (novae). According to present-day
understanding, therefore, less than 10% of lithium could have been
created in the Big Bang, but most of it was produced later in stellar
evolutions from particular nuclear reactions and effects.

From an energy point of view, it is interesting to note that the
evolution of celestial objects (such as the galaxies and stars) is
driven by two main energy sources, gravitational and nuclear, with a
complex interplay of physical processes between them. Gravitational

energy is released during contraction of large masses. Middle-aged *Celestial* stars, like our sun, are powered by nuclear synthesis made possible *energies* by the high temperatures produced by the previous gravitational heating. The supernova explosion, after a preparatory phase of nuclear burning, is triggered by a catastrophic release of gravitational energy, and this makes further nuclear burning possible.

The sun and fossil fuels

The sun is the origin of the energy used up to now to create and maintain life on earth. It is the source of direct solar radiation, wind, and hydroenergy. Fossil fuels also represent an energy source

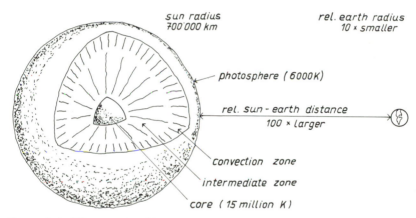

Figure 2.4   The sun provides the energy to maintain life on earth that it has contributed to create.

Its mass is 2×10^{30} kg; its radius 700 000 km (110 earth radii); its average distance from earth 150 million km; its irradiated power 3.8×10^{26} W, derived from the nuclear synthesis within the central core of protons into helium nuclei. This schematic depiction shows the sun's position with regard to earth and its structure up to the photosphere, the "surface" of the sun, which is actually a transition region where density falls off rapidly. The adjacent, outer atmosphere, which is called the chromosphere because it selectively absorbs some small portions of radiation emitted by the photosphere, is relatively transparent and can thus generally be ignored. Not shown is the corona, the very rarefied and hot solar atmosphere of ionized elements that with its complex hydromagnetic structure extends outward by many solar radii (more than 1 000 000 K at densities 10^{12} times smaller than of the earth's atmosphere are attained therein at about two solar radii). (Further information is given in Sec. 4.4.)

created by solar radiation through photosynthesis, the process controlled by chlorophyll, through which plants convert the radiation into chemical energy. The sun, in turn, derives its energy from the nuclear synthesis of protons into helium obtained in the hot central core (Fig. 2.4).

Solar furnace   The solar fuel is thus represented by the protons which have been formed (as we have seen) at the very beginning of the universe, a fraction of a second after its creation. The nuclear fusion process takes place in a central core, at a temperature estimated at about 15 million Kelvin, which is held together by the gravitational forces of the large solar mass. The latter corresponds to 330 000 earth masses and is composed, for more than 96%, of protons and their helium ash.

The proton burning is mainly direct, although the catalyzing carbon cycle might also contribute marginally (reactions 7 to 12 in Ref. 2.1). The energy flows by means of a complex physical process from the central furnace toward the sun's surface, taking about 1 million years to reach it, from where it is irradiated into space and onto the earth.

Photosynthesis   Some 2 to 3 billion years ago, solar radiation contributed in a decisive manner to create and to sustain the first forms of life on our planet, by a process made possible by chlorophyll, which efficiently transforms radiation energy into chemical energy.[2.5] Chlorophyll is a complex chemical substance which, by using most of the solar energy in the visible light spectrum, synthesizes basic plant components, such as cellulose and sugar, out of carbon dioxide and water. In addition, the oxygen released in this photosynthetic activity sustains animal life.

Gondwana and Laurasia   Almost all fossil fuels used today are of biological origin, and the energy canned in them was thus basically derived from solar energy. How this has happened in detail still poses many unanswered questions. Fossil fuels were formed over vast periods of time, particularly (but not solely) around 300 to 350 million years ago. To keep this period in perspective, it may be recalled that this took place before all the emerged land masses, then still assembled into the two adjacent macrocontinents Gondwana and Laurasia, began to drift apart gradually to form the present five continents.

Coal   The prevailing ideas about coal and lignite formation are related mainly to plants growing on the margins of huge marine basins undergoing geological subsidence. The accumulating and partly water-covered dead vegetation was still exposed partially to the atmospheric oxygen. Oxygen reacts more easily with hydrogen than carbon, resulting in a hydrogen depletion of the organic mass. In addition, bacterial action may have contributed to the transformation of wood into coal. The subsiding ground favored the coverage of this mass by sedimentary deposits. In fact, coal is characterized by a ratio of hydrogen to carbon atoms of less than 1, descending to one-hundredth for anthracite, the best quality coal. In addition, coal seams are commonly found sandwiched between sedimentary layers of sandstone, shale, and the like.

On the other hand, petroleum products (the collective designation of crude oil, tar, and natural gas) are characterized by molecules with a hydrogen to carbon ratio of more than 1; specifically, natural gas has more than 2, and the various oils between 5 and 10. It can *Petroleum* be assumed that oil is derived from the decomposition of plant remains (particularly algae) that accumulated in deeper water, where the organic matter was homogeneously intermixed with various sea sediments. This environment would have limited the contact with oxygen (resulting in a larger hydrogen to carbon ratio) and also reduced microbiological activity, thus preserving the organic matter from mineralization, that is, its decomposition into carbon dioxide, water, and nitrogen. The results of this process, organic matter intermixed with inert material, are the various oil shales; as we shall see, oil shales represent energy sources which are orders of magnitude greater in energy content than those represented by pure oil.

Oil is derived from organic deposits in particular circumstances. In fact, the forming shales may have gradually become buried deep and deeper under layers of sediment. Under geothermal heating (normally around 3°C per 100-m depth), they may have gone through *Oil* the "oil window," at about 60 to 150°C where little drops of petroleum *window* condense and eventually migrate into permeable reservoir rocks, like a sponge saturated with oil. An impermeable capping of salt or stony deposits would then allow the accumulation of liquid deposits. In many cases a pocket of gas will develop above the oil reservoir, exerting pressure on the oil below.

As the burial process continues and the temperature increases above 150°C, further decomposition of the oil shale takes place, leading to *Pockets* the formation of natural gas, methane. This is the reason why gas, *of gas* rather than oil, is often found at greater depths, contrary to previous expectations.

According to an exciting but highly speculative hypothesis, huge methane reserves of nonfossil origin could exist at depths of 5 to 10 km beneath the earth's surface; they could be greater by many orders of *Celestial* magnitude than methane reserves of fossil origin.[2.3] If this were true, it *methane* would represent the first really new energy source since the discovery of nuclear energy, thus completely revolutionizing the strategy of energy exploitation on earth. The nonfossil or abiotic hypothesis originates from the finding that carbon (the central component of all organic molecules) is concentrated only in the upper part of the earth's crust. An explanation of this could be that carbon from deep inside the earth, trapped when the solar system coalesced from a cloud of gases and minerals, was carried to the earth's surface in the form of various hydrocarbons. Deep below the surface, these could have been transformed into methane, a stable gas at very high pressures and temperatures, which is forced upward by pressure, unless it is trapped by impermeable layers where it would form the huge new resources.

2.2   ENERGY ON EARTH

Early uses of energy

Fire was probably the first form of energy to be controlled by man, and, in any case, it is closely linked with the evolution of homo sapiens. Fire was being used by man at least half a million years ago. A higher degree of control, which made it possible to light a fire *No safety* whenever needed, dates back to the palaeolithic period of around *match* 30 000 years ago (Table 2.1).

At these early times, fire was used for cooking and heating, a crucial advantage of man over animals, particularly during the ice ages. It is instructive to remember that in the northern hemisphere the last ice *Fire and* age started 70 000 years ago, and after various fluctuations attained its *ice* maximum 18 000 years ago; it then lasted for about 6 000 more years.

Fire was also an element of destruction, and, as such, a weapon in wartime. The dramatic aspects of energy, as symbolized by fire, are well expressed by the Greek myth of Prometheus, who stole fire from heaven and gave it to man, and was therefore put to extreme torture *Prometheus* by Zeus. The Greek poet Hesiod sees in this myth the more tragic *and Zeus* aspects, the punishment for the forbidden act of stealing and giving to man the awful element of fire. On the contrary, Aeschylus sees it rather as heroic act of rebellion against the tyranny of Zeus and as a

Table 2.1 The history of energy; some important dates

15 000 million years ago	● Big Bang, start of evolution of the universe; energy is packed into primordial protons and, later, deuterons.
9 000 to 5 000 million years ago	● Uranium and thorium (later to be found on earth) are synthesized in one or more successive supernova events.
5 000 million years ago	● The sun is formed and starts burning protons.
1 800 million years ago	● First natural fission chain reaction takes place in uranium mine in Gabon.
200 to 300 million years ago	● Major fossil fuel reserves on earth are forming.
30 000 B.C.	● Controlled use of fire.
10 000 B.C.	● Slavery, as testified by Egyptian drawings.
3 000 B.C.	● Water power, this form of energy was to spread throughout Europe in the thirteenth and fourteenth centuries A.D.
1 200 A.D.	● Wind power (3 000 years previously it propelled Phoenician vessels).
1300	● Coal in common use.
1774	● Sunlight concentrated into a furnace used to isolate and characterize oxygen.
1787	● Steam engine, heralding the industrial era.
1856	● First oil-well delivers petroleum in Titusville, U.S., on 27 August.
1882	● Electricity generating station in New York.
1942	● Nuclear fission reactor goes critical in Chicago on 2 December.
1945	● Nuclear explosion at Alamogordo on 16 July using a plutonium test bomb.
1952	● Thermonuclear explosion at Eniwetok atoll on 1 November.
1954	● Launching of the nuclear submarine Nautilus, 21 January, Groton (Connecticut).
1968–1970	● Introduction of competitively priced nuclear energy.

Figure 2.5   An old engraving of Christopher Columbus's Santa Maria. This type of wind-propelled vessel (a nao) was widely used at the beginning of the conquest of the Americas and constitutes one example of the use of wind power.

beneficial gift to mankind, and hence as the liberation of man through the power of science and technology.

The most important aspect of fire in the last millennia is, however, that it has made possible the production of new materials such as clay, *Materials* cement, copper, bronze and iron. These materials created by the intensive use of energy are at the basis of our civilization, and, in fact, periods of evolution are named after them.

Wind was used as a means of propulsion for seagoing vessels in prehistoric times, as shown by Egyptian drawings. Between the second and first centuries B.C., the Phoenicians sailed the length and breadth of the Mediterranean in their ships, plying their trade and spreading their culture. In more recent times, the fourteenth and fifteenth centuries, it was once again the sailing ship, which *Sailing* brought European culture and trade to the East, and thereafter to *west* the Americas (Fig. 2.5).

Wind power has been harnessed in Europe in the form of mechanical energy since the thirteenth century, as we are reminded by Cervantes' idealistic and impractical hero, Don Quixote. Wind power was used much earlier in Persia and in the Far East. Today we are again studying its potential as a major source of energy for the future.

Another fact of history, as significant as it is tragic and with energy-related aspects, is slavery, the origins of which are lost in the mists of prehistoric times, over 10 000 years ago. The Greek historian Herodotus reports that a 100 000 slaves worked simultaneously on the construction of the Cheops pyramid, and their work was so arduous that they had to be "renewed" every 3 months. The con- *Pyramids* struction lasted 10 years. Human energy was used to form, trans- *and* port, and hoist the massive blocks of the pyramid. Much later, the *slaves* unfortunate negro slaves of the American plantations, regarded as human machines, were a key element in the conquest and prosperity of many regions of the United States.

The exploitation of water power was widespread in the East in the closing centuries before the birth of Christ. Very much later, at the end of the eighteenth century, this hydraulic energy contributed to the industrialization of Europe, particularly in certain mountain valleys such as the Swiss Canton of Glarus. Waterfalls and wind are indirect forms of solar energy, having their origin in evaporation and in the differences of temperature and atmospheric pressures.

Even the direct application of solar energy dates back to the pre-

Christian era. A typical example is the salt pan where solar energy evaporates sea-water to leave salt. Legend has it that during the Roman siege of his town of Syracuse in 212 B.C. Archimedes used a system of concave mirrors to concentrate the rays of the sun over *Archimedes* the range "of an arrow," thereby setting fire to the enemy ships. Even if there are ample historical and scientific grounds for concluding that this application of war could hardly have occurred in reality, it shows that the use of solar radiation for setting fire was well known in those times. This was a consequence of the knowledge of the principal laws of geometrical optics (Euclid, 300 B.C.). Two thousand years later the concentration of solar radiation was still *Euclid* used to obtain very high temperatures in furnaces for chemical and metallurgical experiments (Fig. 2.6). For example, Joseph Priestley in 1774 used such a solar-heated furnace to liberate oxygen from mercuric oxide and thus to isolate and characterize this important chemical element for the first time.

The Greek scientist and mathematician Hero of Alexandria is

Figure 2.6   The possibility of concentrating the sun's rays by means of mirrors or lenses in order to reach high temperatures (up to 3000°C have been reached) has always caught the imagination of man, particularly scientists. The photograph shows a large three-lens solar oven developed by the French chemist Lavoisier about 1780. (Courtesy: French Bibliothèque Nationale.)

credited with inventing the aeolipile, the first steam-powered engine, somewhere between 150 B.C. and 150 A.D. It consisted of a hollow sphere free to rotate about a hollow axis through which steam was introduced. The steam was ejected through a pair of bent tubes attached to the sphere's periphery and the reactive force caused the system to rotate. This machine incorporated therefore also the principle of a reactive turbine and of a rocket and demonstrated the transformation of heat into useful work.

Engines and rockets

The industrial revolution

The changes undergone by the organization and the procedures of production in Europe between the end of the eighteenth century and the First World War are commonly known as the "Industrial Revolution." At the root of this revolution is the replacement of manual labor in many productive sectors by machines and also the transformation of thermal into mechanical energy through the steam engine.

Spinning and weaving

The first mechanical spinning machine dates back to 1779, and the first mechanical weaving loom to 1801. The productivity increase obtained with the introduction of machines was remarkable, but it became even more marked as technology progressed. In the case of the textile industry, specifically that of weaving, productivity today is 200 times larger than that obtained with the first mechanical looms, being now about 0.7 workhours per 100 m of tissue; whilst for spinning it is 1 000 times larger, i.e., about 0.06 workhours per 1 kg of yarn.

The steam engine was perfected by Papin in 1690, and subsequently by Watt in 1782. In 1787 the first steam engine in the textile industry came into service. In the present century, the twin-piston machine has been replaced by the steam turbine invented in 1884 by Parsons. As mentioned in Chap. 1, the advent of the steam engine induced some fundamental socioeconomic consequences; in particular, it revolutionized transport (Fig. 2.7).

Energy and revolution

The use of coal in Europe goes back as far as 1300 A.D. Coal has the important property of burning with a hotter flame than charcoal or wood. The use of coal made possible the large-scale production of iron from iron ore. In the eighteenth century, iron became the basic metal for constructing machines, since it was cheaper than

Coal and iron bronze. The combination of coal and iron was thus the basic ingredient of the industrial revolution. Coal also acquired an important role as a primary energy source, in particular for providing the thermal energy to drive the steam engine. By means of a distillation process, a practical energy carrier, gas, was also obtained from coal (now substituted by natural gas).

In 1856 the first drilled oil-well delivered petroleum in Titusville, Pennsylvania. Upwelling oil, usually intermixed with water, was known previously, and at the beginning of the eighteenth century in

Figure 2.7   Car driven by a steam engine, built in 1769 by Cugnot.

the United States it was used as a medicine and later as an occasional
substitute for whale oil; then it was the basic fuel for oil lamps. *Oil and*
However, after Titusville the situation changed rapidly: 650 000 *whales*
barrels of oil produced in 1860 increased to over 3 million in 1862.
The Standard Oil company of John D. Rockefeller dominated the
market, mainly fed by oil from Pennsylvania and California, until in
1911 it was forced by antitrust legislation to split into the first three
of the "Seven Sisters": Exxon, Mobil, and Socal. As oil in Texas *The seven*
began to flow at the beginning of this century, two new American *Sisters*
companies were added: Gulf and Texaco. At about the same time
two European companies joined the exclusive club: Shell, initially
exploiting Caucasian and Indonesian oil, and BP, dominating the
Near East, particularly the Persian region.

The history of the petroleum industry is a continuously upwinding
spiral. The use of petroleum started to erode the coal market, since *An upwinding*
it was cheaper to extract and more practical to use. From about *spiral*
1920 onward, the internal combustion (explosion) engine, the basis
of the automobile industry, began to constitute a large and exclusive
market for petroleum.

Discoveries in physics at the beginning of the nineteenth century
paved the way for the generation and exploitation of electricity.
Electricity is, of course, not a primary energy source in the sense of
coal or oil. It is simply a very convenient form of energy, easily
transported and used, which is generated artificially, mainly from
thermal or hydroelectric energy.

An electric motor was first fitted to a boat in 1834, and in 1879 the *Dream*
first railway vehicle was propelled by electrical power in Berlin. *boat*
Three years later Edison built the first electric power station to
supply lighting to part of the City of New York. This marks the start
of a new evolution, since the availability of electricity introduced
increasingly sophisticated and efficient applications of energy in
industry, in the transport sector, and in communications.

2.3 ENERGY IN THE ATOM

From Greek philosophers to neutrons

Nuclear energy together with gravitational energy represents, as we
have seen, the basic energy store throughout the genesis and the

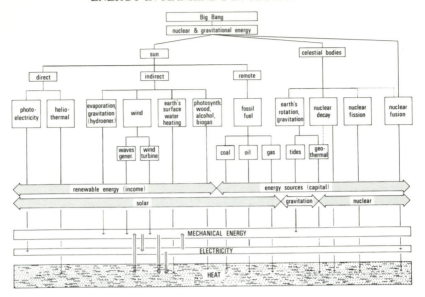

Figure 2.8   Origin and transformations of energy used by man.

All energies derive from the nuclear and gravitational form available once the primordial Big Bang (thermal) energy had partly condensed into matter (proton, neutrons, deuterons, electrons etc.) more than 15 billion years ago. The chart shows the principal information only. Most of the cross connections are not shown, for example, synthetic fuel production by application of solar and nuclear primary energy. There also could be direct electricity production by chemical direct conversion, as in fuel cells and batteries.

evolution of the universe. In fact, practically all energy available to man on earth is related directly to nuclear energy, since solar radiation also derives its energy from the nuclear fusion process (Fig. 2.8).

Small and amazing   How man was able to tame this energy on a practical scale is an amazing story, which we shall recount briefly in this and the following section. It is the story of nuclear physics, of how man was able to discover and probe the infinitesimally small nucleus buried deep in the atom and packed with an awfully large amount of energy[2.6]—in fact, a million times more energy than what can be obtained from the atomic structure.

Physics, the study of nature, was born about 600 B.C. within the Ionian Greek culture as an offspring of philosophy, and can be

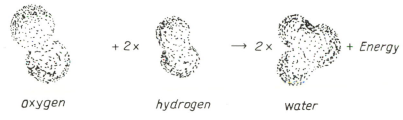

oxygen   hydrogen   water

Figure 2.9   Reaction of a molecule of oxygen (O_2) and two of hydrogen (H_2) to form two molecules of water (H_2O) and release a net chemical combustion energy of 3 eV per molecule, or 15.9 MJ per kg of water.

personified by Thales of Miletus, who, in a rather vague way, created both the concept of "matter" and the principle of conservation of matter. Greek physics (the philosophy of nature) reached its full development within three centuries, with the creation of the concept of the atom by Leucippus and Democritus. *Democritus*

In the middle of the nineteenth century, physicists such as Joule, Kelvin, Maxwell, Clausius, and Boltzmann developed the idea of the "kinetic," moving atom as the nature of heat. This, by conversion, helped to concretize the concept of the atom that was then further refined by chemists and spectroscopists. In this way, without any knowledge of the atomic structure, the periodic table of the atomic elements, which describes the chemical properties of the 92 natural elements, was already known at the beginning of this century. Chemistry explains how atoms unite in groups to form molecules, *Atom* e.g., how two atoms of hydrogen link up with one atom of oxygen to form the water molecule, thus releasing combustion energy (Fig. 2.9). In this period, physics was enriched by fundamentally new concepts such as the energy quantum of Planck (1900); the equivalence of matter and energy of Einstein (1905); the atom model of Rutherford and Bohr (1911−1913).

Toward the end of the 1920s, nuclear physics expanded from the consistent ground established by quantum theory after the First World War. It came to a first completion with the discovery of the neutron by Chadwick in 1932. In fact, with the neutron, the basic concept of the atom was established whereby electrons orbit around *Neutron* the nucleus which is itself composed of protons and neutrons (Fig. 2.10). In this same year, deuterium, or heavy hydrogen, was found by Urey, and a peculiar new particle that has no mass at rest, called the neutrino, took shape and became a fundamental element

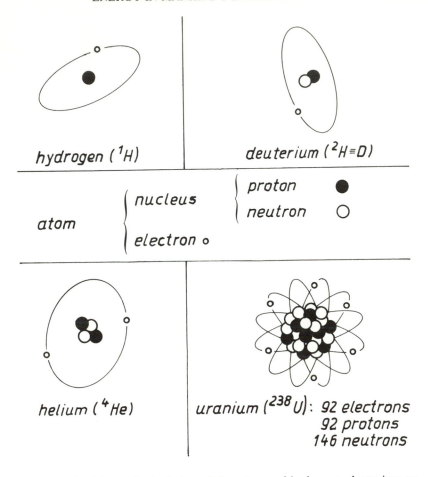

hydrogen (^1H)   *deuterium ($^2H \equiv D$)*

atom $\begin{cases} nucleus \begin{cases} proton & \bullet \\ neutron & \circ \end{cases} \\ electron \; \circ \end{cases}$

helium (^4He)   *uranium (^{238}U): 92 electrons*
92 protons
146 neutrons

Figure 2.10 Schematic depiction of the atoms of hydrogen, deuterium or heavy hydrogen, helium, and uranium, made up of electrons, protons, and neutrons.

In actual fact the radius of an atom is of order 10^{-10} m, and that of its nucleus is of order 10^{-15} m; whereas the mass of the proton is 1.67×10^{-27} kg, the neutron's mass is only very slightly larger, and the electron's mass is nearly $2\,000$ times smaller, 9.1×10^{-31} kg. The isotopes of an element contain the same number of protons (and hence electrons) but a different number of neutrons. For example, the depicted uranium 238 and uranium 235 are two uranium isotopes, the latter also with 92 protons and 92 electrons, but only 143 neutrons; hydrogen and deuterium are also isotopes. An isotope is characterized by the element's name and the sum of the protons' and neutrons' numbers, e.g., written as helium 4 or ^{4}He.

in nuclear physics; it had first been proposed by Pauli in 1930.

The nucleus of deuterium, the deuteron, made of the simple combination of a proton and a neutron, played a central role in developing nuclear physics in the 1930s in much the same way as the hydrogen atom had in atomic physics in the previous decade. Man found in the deuteron the nuclear energy package that had been *Energy* created, as we have seen, more than 15 billion years ago. *package*

The 1930s can be characterized as the great decade of experimental nuclear physics. It was demonstrated that artificial reactions, particularly among light nuclei and those induced by neutrons, could be produced in the laboratory and that energy was released in them. The exothermic reaction of two deuterium nuclei resulting in a tritium nucleus and a proton (one of the basic nuclear fusion reactions) was described by Rutherford and collaborators at the Cavendish Laboratory in 1934.

A fundamental question of astrophysics, namely, about the energy source of the sun and stars, was connected to nuclear energy. As early as 1929, Atkinson and Houtermans clearly postulated the concept of thermonuclear fusion reactions providing the energy radiated by the stars and the sun. Refined work on this subject by many physicists (for example, Gamov and Teller, von Weizsäcker, *Stellar* Bethe) and the many new results obtained in the field of nuclear *furnace* physics clearly indicated that the synthesis of protons into alpha particles—through various reaction channels—was the basic process of the stellar and solar furnaces.

The concept of thermonuclear fusion as a powerful nuclear energy source was thus well established by 1937. Today it may seem somewhat surprising that no suggestions for achieving a practical fusion energy source on a terrestial scale were seriously considered before World War II (this subject was to be taken up after the war, as we *Fusion* shall discuss in Chap. 5). Actually, no less than Rutherford said in *and* 1933 about speculations on power generation from nuclear (fusion) *moonshine* reactions: "Anyone who expects a source of power from the transformation of these atoms is talking moonshine." But then, from the end of 1938 onward physicists turned their attention to the exciting new results from fission experiments.

The advent of nuclear fission

Crisscrossing fission The breakthroughs, which actually gave rise to the practical exploitation of nuclear energy, came around 1939 and concerned not fusion but the nuclear fission process. As is usual for important breakthroughs, there was no single discovery that showed how fission could produce nuclear energy at levels of practical interest, but rather a crisscrossing sequence of discoveries made at various times. If we attach a few names to these discoveries, it is as in most other cases only to mark time periods or to characterize groups of people, and not because they have exclusive credits.

In a way, the final discovery of fission grows out from the experimental demonstration in 1934–1935 that artificial radioactivity can be produced with alpha particles (Joliot and Curie in Paris), or with neutrons (Fermi in Rome). Both these groups, and about one year later Scherrer in Zurich[2.4] and a graduate student in Berkeley *A war* (Abelson), narrowly missed understanding and explaining the fission *not* process. If nuclear fission had been discovered in the mid-1930s, the *missed* historical events leading to World War II would have probably changed completely. How? It is difficult to say: Probably war could have been prevented altogether, since the opposing blocs would have certainly feared the devastating armageddon emerging from the new discovery.

That neutrons can cause uranium to fission was discovered in December 1938 by Hahn and Strassmann in Berlin; Frisch in Denmark and Lise Meitner in Sweden correctly interpreted the discovery by demonstrating the great release of energy that follows from fission (Fig. 2.11). A rapid sequence of discoveries from March 1939 to Spring 1940 hinted at the possibility of starting a nuclear chain reaction (speculations about this possibility had been nourished by Szilard since 1935). In particular, around this period it was found that neutrons are produced during fission, possibly two or three per fission event (Anderson, Fermi, and Hanstein, U.S.; Szilard and Zinn, U.S.; Halban, Joliot, and Kowarski, France); *Chains* that uranium 235 is the fissionable isotope of uranium (Bohr and *released* Wheeler, U.S.); that a self-sustaining nuclear reactor could be built if a suitable moderator were to be found; and that carbon is indicated as such a moderator (Anderson and Fermi, U.S.). Then, on 2 December, 1942, under the supervision of Fermi, the first nuclear reactor manufactured by man became operational in Chicago. The nuclear age had begun.

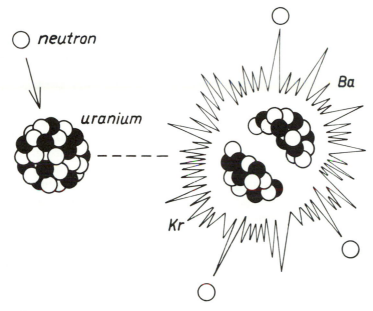

neutron

uranium

Ba

Kr

Figure 2.11 Exothermic nuclear reaction. Fission of the uranium nucleus subsequent to neutron capture, into two intermediate nuclei (e.g. barium and krypton) releasing 2 or 3 neutrons and 200 MeV (million electronvolt) of energy.

These scientists did not then know that a natural chain reaction had already taken place in the uranium mine of Oklo in Gabon some 1800 million years previously. It was not until 1972 that French scientists discovered the traces of at least six fossil nuclear "reactors"; these became critical through the infiltration of water into the mine, consuming a total of over 500 tons of uranium in the space of a million years.[2.8]

Reactors in Africa

The Chicago pile, more a physical experiment than a reactor, was operated in great secrecy, as it was part of the gigantic U.S. effort to build an atomic bomb, the operation being called the Manhattan project. A reactor was necessary to breed the artificial but fissile plutonium needed to manufacture one type of atomic bomb. The feasibility of this production was demonstrated in the second reactor, the one at Oak Ridge, which is still preserved as a historical landmark.

Atoms in Manhattan

In the public eye, the nuclear age was brutally heralded by the explosion of the atomic bomb dropped on Hiroshima in 1945. This was not, however, the first nuclear explosion, since a few months earlier a test detonation had taken place in Alamogordo in the New Mexico desert using a plutonium implosion device.

It is evident from these facts that the advent of nuclear energy received a substantial boost from the war effort. Brought tragically to the limelight by the bombs dropped on Hiroshima and Nagasaki, this event affected and continues to affect adversely the acceptance of nuclear energy in peaceful applications. Unfortunately, the big- *Explosive* gest drive in the field of nuclear research in the immediate postwar *on our* period was not to be for peaceful applications, but once again in a *heads* military context as part of the arms race, where nuclear fission bombs were perfected and the nuclear fusion bomb was developed. The result of this incredible race is the existence of an enormous arsenal of nuclear bombs, which today amount to a yield of several tons of equivalent chemical explosive per inhabitant of our planet.

Nuclear reactors were also developed for military applications, first for plutonium production and later also for the propulsion of nuclear submarines. This latter application—developed under the *Energetic* striving auspices of the energetic Admiral Rickover, "father" of the *submarine* large nuclear-powered U.S. fleet—at last gave nuclear energy a decisive push toward peaceful applications. The present generation of relatively compact nuclear reactors with enriched uranium fuel and light water moderators is in fact a direct spin-off from the options taken at the beginning of the 1950s for submarine propulsion.

In conclusion, it is interesting to note that if nuclear fission has today become a competitive, primary energy source, this is due in a *Heavenly* way to a heavenly blessing. In fact, there is, and was, no theoretical *blessing* reason why the number of neutrons emitted per fission of the uranium 235 isotope (the only natural isotope available on earth that is fissionable with thermal neutrons) should be sufficient to maintain the chain reaction, which has made this energy source possible.

As described in this chapter, the nuclear success story began with fusion, which produced (and still produces) energy and the synthesis of nuclei in the universe. But fission was the first to happen on earth, in the Oklo uranium mine 1.8 billion years ago.

Man was led into discovering nuclear energy in the 1930s by studying thermonuclear fusion, but it was fission which first produced nuclear energy on a practical level on earth, both in a controlled (Chicago, 1942) and an explosive way (Alamogordo, 1945). Then, attention was recaptured by a fusion event—the first detonation of a thermonuclear fusion device at the Eniwetok Atoll in 1952—which has since greatly influenced and changed political and *Detonating* military strategies and international relations through the awful *atoll* existence of fusion weapons. Shortly thereafter, fission was brought into focus again by becoming a truly competitive new energy source, which, in its technologically most advanced version (the breeder), could contribute to the energy needs for centuries to come. So, it would now seem to be fusion's turn again, a subject that we shall discuss in Chap. 5.

CHAPTER 3

Nuclear Fission Energy

Nuclear energy, one of the two fundamental energy sources that drive the evolution of the universe, represents the stage of energy development nearest to the present time and perhaps the most complex one. The public's reaction to nuclear energy is still partly conditioned by its advent in a military context, and by incomprehension. It remains a fact that, for the broad public, the atomic mushroom has for years been the symbol of the destructive power of certain scientific and technological discoveries. Will nuclear energy resume its forward march and sooner or later become one of man's predominant energy sources, as has so often been predicted in the past?

3.1   THE FISSION SOURCE

Nuclear electricity today

In spite of all discussions, nuclear energy is already used on a wide scale. At the end of 1984, there were 344 nuclear reactors for electricity production in service in 24 countries, providing an installed electronuclear capacity of 220 GWe (billion watts of electric power); another 210 reactors, either under construction (180) or on firm order, will about double the existing capacity. In 1984 the world produced more than 13% of its electricity from nuclear energy (13.5% in the United States and 23.6% in the European Community), and with the reactors presently under construction, the share will increase to about 17% in 1990.

Projections for the year 2 000 made in 1981 indicate that the total installed electronuclear capacity will be about 500 GWe for the *Wrong projection* OECD countries (including essentially the United States, Canada, Western Europe, Turkey, Australia, Japan). This figure is about a quarter of what was predicted as recently as 5 years earlier (and may

56

be too high once again), and clearly demonstrates the slowdown of the application of nuclear energy in the last few years. The Soviet Union, on the other hand, has announced plans for a joint nuclear program with its Comecon partners. In addition to the already operational 16 GWe, the program aims to install some 30 GWe of new capacity by 1990. China plans to have up to 15 GWe of nuclear power in service by the end of the century.

Beneath this global picture of slow but nevertheless steady expansion lie some widely different attitudes and prospects with regard to nuclear programs. Whereas France, Belgium, Finland, Taiwan, *Nuclear* Bulgaria, Sweden, and Switzerland, for example, had a nuclear *age* contribution to electricity production of between 60 and 30% in 1984 (with further increases due in the coming years), West Germany, in spite of its important nuclear industry, has stayed with a contribution of about 23% in the face of violent opposition. This, however, is still more than the Italian contribution of less than 4%. On the other hand, six U.S. states have shares of more than 35%, with Vermont reaching the 80% level.

The overall situation would seem to indicate that a majority of people accept nuclear energy and favor its further extension within *Reasonable* reasonable limits. Nevertheless, a concerned and active minority *limits* questions the merits of the nuclear option on the basis of one or all of three major issues:

- Safety of reactors and of fuel cycles
- Long-term disposal of radioactive waste
- Proliferation of nuclear weapons

Sooner or later, however, even the nuclear critics will be forced to weigh up in comparable terms the advantages and risks of the various competing forms of known energy sources. Once there is a clearer realization of the limited dimensions of some of the alternative sources, or of the environmental effects of burning fossil fuels—probably the greatest ecological danger of them all—nuclear power will most likely continue its forward march.

The reactor

Nuclear fission has become an energy source of practical importance as a consequence of the miraculous existence on earth of the natural fissile uranium 235 element, which is able to sustain a propagating

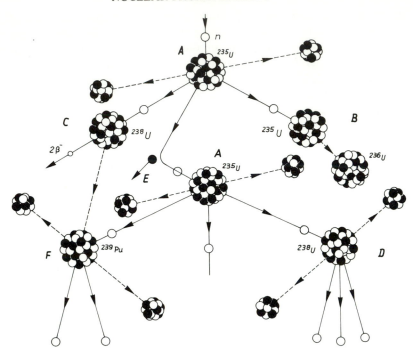

Figure 3.1 Schematic presentation of some of the main nuclear reactions which can occur in a uranium reactor.

In actual fact, each one of the reactions described occurs with different degrees of probability.

A: Neutron capture by uranium 235 isotope with subsequent fission of the nucleus into two radioactive fragments and with emission of three secondary neutrons (about 85% of the thermal neutrons captured will cause fission).

B: As in A, but with a different pattern of reaction, i.e. formation of uranium 236 (which, with a half-life of 24 million years, decays into thorium 232 by α particle emission).

C: Neutron capture by uranium 238 resulting in the formation of fissile plutonium 239 by β particle emission.

D: As in C, but with a different reaction pattern, i.e. fission of uranium 238 into two intermediate nuclei with release of three neutrons.

E: Reflection of a neutron through impact with a light nucleus, e.g. the hydrogen nucleus, with transfer of part of the kinetic energy of the neutron to the nucleus ("slowing-down" of the neutron).

F: Fission of plutonium 239 into two intermediate nuclei with emission of two neutrons (about 65% of the thermal neutrons captured will cause fission).

chain reaction as it emits several neutrons by the fission of its
nucleus per neutron absorbed. As a consequence, the abundant and *Miraculous*
fertile uranium 238 and thorium 232 elements can also be trans- *uranium*
formed into fissile elements (plutonium 239 and uranium 233) by
capturing the excess neutrons produced, in the first place, by the
uranium 235 fission process.

The understanding of how a nuclear reactor works, that is, how
uranium is fissioned under controlled conditions, requires a knowledge of the neutron's basic properties as it moves and eventually
is absorbed in the reactor core composed of uranium and inert
materials. Some of these properties are summarised qualitatively in
Fig. 3.1, and more details are given in Ref. 3.1.

After being emitted through the fission process, the neutrons are
slowed down by collisions with other nuclei (unless they are absorbed
by them), just as occurs in the game of bowls. This process of
moderation is, in fact, necessary to operate most present-day uranium
reactors, since fission of uranium 235 is enhanced if the neutrons are
relatively slow (moreover, slow neutrons are less easily absorbed
into nonfissile uranium 238, which is 150 times as plentiful as fissile *Bowl's*
uranium 235). The moderation effect is artificially obtained by *game*
introducing light elements into the reactor, of which the most
common are hydrogen and deuterium (in the form of normal or
heavy water) and graphite.

Practically all civilian reactor fuels are in the chemical form of
oxides, whereby the uranium is nearly always enriched in the fissile
uranium 235 isotope from its natural abundancy of 0.7% up to
between 2 and 3.5%.

Besides the nuclear fuel and (where necessary) the moderator, a
reactor includes the control rods as a basic component. These contain neutron absorbing elements, and thus the neutron population in
the chain reaction—and as a consequence, the power production—
can be regulated by the variable rod position within the reactor
(Fig. 3.2). Reactor control is relatively simple, because the emission
of nearly 1% of the generated neutrons is delayed for intervals *Tamed*
ranging from a fraction of a second to more than a minute after the *neutrons*
fission event, these delayed neutrons being emitted by the fission
products. In the normal operating mode of a reactor, the prompt
neutrons alone cannot sustain the chain reaction, but the combination
of prompt and delayed neutrons can do so. This then introduces a

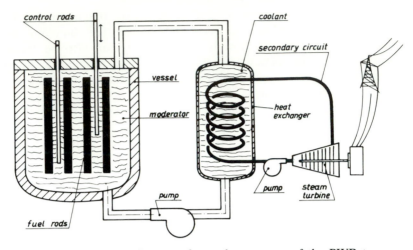

Figure 3.2   Schematic diagram of a nuclear reactor of the PWR type.

time constant for control, which can be easily handled by modern servomechanisms.

Finally, a coolant transports the heat from the nuclear reactor core into a conventional thermal cycle where, eventually, it drives a vapor turbine and produces electricity.

Combinations of various types of fuels, moderators, and coolants *Pressurized* give rise to a large variety of different reactor concepts, many of *boiler* which have been tested or are operational today (Table 3.1). By far the most common civilian reactor type is the pressurized water reactor (PWR), which uses normal water both as a moderator and coolant and has its fuel enriched to about 3.3%. Presently, for each three PWRs, there are about one boiling water reactor (BWR) and one of any other type in operation and construction, with the PWRs further increasing their share in the future.

The core of a 1.2-GWe pressurized reactor, containing about 100 tons of enriched uranium oxide, is enclosed in a steel pressure vessel, typically 5 m in diameter and 15 m high, with walls 15 to 30 cm thick. The core is cooled through at least two independent primary loops containing water at 320°C under pressure of 170 atm so as to prevent boiling; a heat exchanger generates steam at 285°C and 70 atm in a secondary loop to drive the turbines. Annual fuel consumption and waste production are given in Table 3.2. The PWR type was originally developed because of its compactness for

Table 3.1 Some nuclear reactor types

Type	Fuel		Moderator	Coolant Primary * Secondary	Remarks
	Composition	Typical enrichment %			
LWR—light water reactor	UO_2	2.5–3.5, ^{235}U	H_2O	H_2O * H_2O	By far the most common civilian type, with the PWR and BWR as subclasses
PWR—pressurized water reactor	UO_2	3.2, ^{235}U	H_2O	H_2O * H_2O	Compact, most common type, in which the primary water coolant is kept from boiling by containment under high pressure (170 atm, 320°C)
BWR—boiling water reactor	UO_2	2.8, ^{235}U	H_2O	H_2O * None	The vapor formed in the reactor goes directly to the turbines at 70 atm, 300°C, as compared with 70 atm, 285°C for a PWR, and 40 atm, 250°C for a CANDU
HWR—heavy water reactor	UO_2	Natural	D_2O	D_2O * H_2O	The high neutron effectiveness of heavy water as a moderator (in typical quantities of 1 t per MW of electrical power) allows the use of natural uranium as fuel

Table 3.1 (Continued)

Type	Fuel		Moderator	Coolant	Remarks
	Composition	Typical enrichment %		Primary * Secondary	
CANDU—Canadian deuterium uranium reactor	UO_2	Natural	D_2O	D_2O * H_2O	The fuel contained in hundreds of horizontal tubes with the coolant at 100 atm, 310°C, can be loaded at full power
GCR—gas-cooled reactor	UO_2 (+ThO_2)	2–3, ^{235}U (2–3, ^{233}U)	Graphite	Gas * H_2O	Because of graphite moderator (typically 700 t per GWth) it has good breeding capabilities and can be thus characterized as converter
HTGR—high-temperature gas-cooled reactor	UO_2	2.5, ^{235}U	Graphite	He * H_2O	Operating temperatures above 1000°C are in principle possible. At demonstration plant of Fort St. Vrain helium is at 70 atm, 690–820°C, and steam at turbines at 170 atm, 540°C
AVR—Arbeitsgemeinschaft Versuchsreaktor	UO_2 (+ThO_2)	2.5, ^{235}U (2, ^{233}U)	Graphite	He * H_2O	Graphite spheres, the size of tennis balls, containing the nuclear fuel "flow" slowly through the "fluidized-bed" type core where they are cooled by helium at more than 900°C
AGR—Advanced gas-cooled reactor	UO_2	2.3, ^{235}U	Graphite	CO_2 * H_2O	British commercial reactor type, developed from the previous MAGNOX type (that uses metallic natural uranium), which itself derives from plutonium-producing reactors

Reactor	Fuel	Fissile	Moderator	Coolant	Comments
BR—breeder reactor	UO₂(UC) ThO₂	6–20, ^{239}Pu 0.3–5, ^{233}U	Various	Various * Various	Produces more fissile elements (^{239}Pu, ^{233}U) than it consumes; as this requires generally a "faster" neutron spectrum than in thermal reactors, it has no or only a modest moderator
LMFBR—liquid matal fast breeder reactor	UO₂+ +PuO₂	20, ^{239}Pu	None	Na * Na	By far the most studied and available breeder has a compact core and no moderator; water vapor is produced in a tertiary coolant loop
MSBR—molten salt breeder reactor	Li, Be, Th, U fluorides	O.3^{233}U	Graphite (LiF+BeF)	Fuel (salt) * Salt	Most fascinating breeder reactor of the quasi-thermal type, where fission, breeding, heat removal, fuel loading, and reprocessing are all made in or by a mixture of lithium, beryllium, thorium, and uranium fluorides; this molten salt flows continuously from the core, where it is made critical by a graphite moderator, to the heat exchanger and reprocessing unit. Extreme corrosion problems have prevented up to now operation of a test reactor

Table 3.2   Nuclear waste of a 1-GWe electric power station with the PWR-type nuclear reactor[3.2] (containing 82 t of uranium initially enriched to 3.3% in uranium 235).

Annual consumption[3.3]

690 kg	uranium 235
650 kg	uranium 238

Mean annual production[3.4]

960 kg	Waste, fission products
240 kg	Plutonium (of which 170 kg fissile isotopes); an additional 360 kg of plutonium are produced and consumed directly in the fuel } Transmutation products
140 kg	Nonfissile elements (actinides)
21 700 GWh	Heat (equivalent to 870 g of matter) of which one-third is converted into electricity

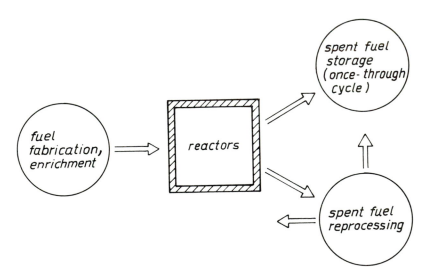

Figure 3.3   The nuclear reactor and its fuel cycle.

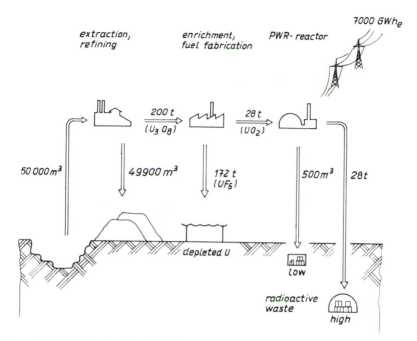

Figure 3.4   Fuel cycle of a 1-GWe pressurized light water reactor (PWR), with no reprocessing of spent fuel (indication of annual quantities involved). (Reprocessing would recover for reuse in other reactors 26.6 t of uranium and 170 kg of plutonium.)

use in nuclear submarines (where, actually, to further reduce balk, the fuel is enriched to more than 90%).

It is important to realize that the technical and commercial success of the nuclear power generation process depends not only on the reactor but also on its related fuel cycle; this is particularly true for future and advanced reactors, such as the fast breeder. The fuel cycle determines reactor operation, as it lies both before and after it (Fig. 3.3). Fuel fabrication is well developed, being obviously a *Fuel's* necessary element for reactor operation and having gained substan- *cycle* tial advantage from the military interest in nuclear fission. Fuel reprocessing (and storage), on the contrary, has benefitted only partly from this advantage and, in fact, is not yet operational on a substantial level.

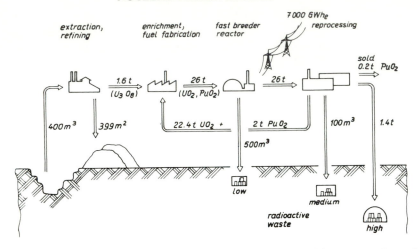

Figure 3.5   Idealized fast breeder reactor fuel cycle with reprocessing of spent fuel to ensure the production and rapid recycling of the relatively large plutonium amounts.

The reported annual quantities would be obtained only in a near steady-state operation of hundreds of fast reactors after tens of operation years, and they assume that 60% of the extracted and produced uranium is finally burned up for energy production.

3.2   FUEL CYCLE

The fuel cycle comprises all the operations carried out on the nuclear fuel, from the extraction of uranium or thorium ores to storage of the radioactive waste (Figs 3.4 and 3.5). It represents a substantial cost component of nuclear energy and requires huge investments.

Fission fuel

The fission fuel of present-day reactors is almost exclusively uranium; it derives its potential energy either directly from the only fissile isotope in nature, uranium 235, or from the artificial plutonium 239 *World* produced through neutron capture in uranium 238. (Plutonium fuel *uranium* is here, so to speak, a waste product and will thus be discussed in the next section.) Approximately 70% of the world uranium requirement is at present produced by Australia, the United States, and Canada. Potential major suppliers for the future are, for example, the Soviet Union, Chad, Namibia, Nigeria, Somalia, and China. The cost of uranium rose from $9/kg in 1965 to roughly $100/kg in 1981.

The enrichment of uranium with its fissile isotope uranium 235, from its natural concentration of 0.7% to around 3%, is a major

Figure 3.6   Separation of uranium 235.

The early (military) applications of nuclear energy necessitated the generation and accumulation of fissile nuclei, either by transmutation (plutonium 239, formed in reactors) or by separation of the fissile isotope (uranium 235) from natural uranium. The photograph shows the first separation installation, which started production at the beginning of 1944 at Oak Ridge, Tennessee. Separation was achieved electromagnetically by means of the huge magnet shaped like a racetrack. Because of the scarcity of copper in the war years, electric cables made of silver were used in the coils of the magnet: 15 000 tons of silver lent by the Treasury (and subsequently recovered) were used in this installation, in which production soon ended in 1945 when enrichment by diffusion proved more efficient.

operation in the preparation of the nuclear fuel required by light
water reactors. Enrichment presupposes the capability of separating
the isotope uranium 235 from the more abundant uranium 238. On

Enrichment an industrial scale this is done by diffusion or centrifuge methods of
by the chemical compound UF_6-uranium hexafluoride, commonly
separation known as "hex." On the laboratory scale, or in pilot plants, separation is also obtained by means of supersonic nozzles, electromagnets, lasers, or by exploiting special effects in chemical reaction
dynamics or in plasma rotation. The laser separation method looks
promising and has been chosen by the U.S. Department of Energy
as the method for the future. Historically, the first method applied
in 1944 on a large scale was the electromagnetic mass separation
obtained in the huge "Calutron" magnet units at Oak Ridge
(Fig. 3.6).

The diffusion method is by far the most widespread and is based
on the slightly higher diffusion rate of uranium 235, in relation to
the heavier uranium 238, through porous ceramic membranes or

Figure 3.7   Diffusion enrichment installation.
The two parallel structures in the center of the photograph show the first
operational installation for the enrichment of uranium 235 using the diffusion
method; production began at the end of 1944 at Oak Ridge, Tennessee.
The plant (1980) produced enriched uranium for approximately 50 nuclear
power stations but is now closed down.

"barriers." The compressors, which pump the gaseous hex around
the gigantic diffusion installations comprising thousands of stages, *Porous*
consume an enormous amount of energy. Of the electricity pro- *barriers*
duced by light water reactor power stations, 4% is used to enrich
the fuel which they consume.

Large diffusion plants are in operation in at least six locations
around the world (Fig. 3.7): two in the United States (Portsmouth,
Ohio; Paducah, Kentucky); one each in France (Pierrelatte/Tricastin),
and in the United Kingdom (Capenhurst); and at least one each in
the Soviet Union and China.

The centrifuge method is about 10 times less energy-consuming
and allows the use of smaller plant units as compared with the
diffusion method.[3.8] Production of enriched uranium in demonstration (or production) plants is underway in various locations, for
example, in Capenhurst (U.K.), in Almelo (The Netherlands). *Centrifuges*

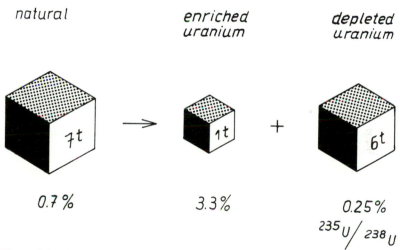

Figure 3.8   Enriched and depleted uranium.
  The enrichment process is often measured in SWU (separative work
units) the quantity of which depends on the rate of enrichment and of
depletion; the example illustrated here corresponds to approximately 4 500
SWU, i.e., it takes 4.5 SWU to produce 1 kg of uranium enriched to 3.3%
with a tails assay of 0.25% (the cost is about $100 per SWU, or, alter-
natively, $35 000 per kilogram of uranium 235).

Approximately 6 tons of depleted uranium is left over for every ton of enriched uranium which goes to the light water reactors (Fig. 3.8). By 1982 the worldwide production of enriched uranium for civilian and military applications probably left over more than *Energy* 400 000 tons of depleted uranium with a mean uranium 235 content *bonanza* of around 0.25%;[3.9] it is generally stored in the form of hex (UF_6) in stainless steel barrels near the separation plants.

With a potential total energy content of 1.76 tep per gram of uranium (see Table 4.1), the stored 400 000 tons of depleted uranium corresponds to 7×10^{11} tep (ton of petroleum equivalent), or approximately the worldwide proven and estimated oil reserves! The efficient exploitation of this enormous energy reserve (more than $50 000 million worth) is possible through the fast breeder reactor which breeds fissile plutonium from uranium 238. Another alternative for a very partial energy recovery from this reserve would be provided by a very efficient and selective separation method, which allows the extraction of the remaining 0.25% of the fissile uranium 235 (in this respect, some attention is paid to research on laser-assisted separation methods).

The other important nuclear fission fuel is plutonium, as mentioned previously. It must be stressed that producing and burning plutonium has been made possible, in the first place, by the existence in nature *Magic* of the fissile uranium 235. This isotope has allowed the fission chain *touch* reaction to start, and thereby—by a sort of magic touch—makes it possible to gradually transform the large uranium 238 (or, alternatively, thorium 232) reserves into the fissile plutonium 239 (uranium 233), as will be discussed in more detail in the next chapter.

Radioactive waste

Any discussion on the future of nuclear energy ultimately converges on the real and presumed problems represented by the radioactive waste, which is made up of fission and transmutation products. The latter result from the capture of one neutron by uranium or by the subsequently formed nuclei, which after some transformations lead to various transuranic elements (elements with atomic numbers greater than that of uranium; see Table 3.3).

Plutonium 239 is the most important of these elements since it is also a basic nuclear fuel. Other man-made elements have found

Table 3.3   Properties of some radioactive nuclei*

Element	Atomic or Z number	Isotope Neutron number	Isotope Mass number and symbol	Decay time (years)	Remarks
Natural actinides (mainly α emission)					
Thorium	90	142	^{232}Th	14 000 million	
Uranium	92	143	^{235}U	710 million	Fissile
Uranium	92	146	^{238}U	4 500 million	α decay followed by 85 keV γ rad.
Artificial elements (mainly α emission)					
Uranium	92	141	^{233}U	160 000	Fissile
Neptunium	93	144	^{237}Np	2.1 million	
Plutonium	94	144	^{238}Pu	86	0.57 W/g decay heat
Plutonium	94	145	^{239}Pu	24 000	Fissile
Plutonium	94	146	^{240}Pu	6 800	Fissile; spontaneous fission, with 1 000 neutr./g·s
Plutonium	94	147	^{241}Pu	14	Fissile (β decay)
Plutonium	94	148	^{242}Pu	400 000	
Americium	95	146	^{241}Am	433	
Americium	95	148	^{243}Am	7 000	
Curium	96	148	^{244}Cm	18	
Nuclei from fission (β and γ emission)					
Strontium	38	52	^{90}Sr	29	
Zirconium	40	53	^{93}Zr	900 000	
Technetium	43	56	^{99}Tc	200 000	
Iodine	53	78	^{131}I	0.022	
Cesium	55	82	^{137}Cs	30	
Krypton	36	49	^{85}Kr	(4.4 h) 9.4	gaseous

*Types of nuclear "radiation":

Alpha (α). Consists of helium nuclei moving at high velocities; absorbed by a sheet of paper.
Beta (β). Consists of electrons moving at very high velocities; absorbed by a sheet of metal.
Gamma (γ). Consists of electromagnetic radiation at extremely short wavelengths; most penetrating, markedly absorbed by ½ m of concrete.

Friendly applications in industry, medicine, and communications. Some of
decay them, as plutonium 238, are used in power generators where electricity is produced from their nuclear decay; such generators are
being used to power devices for which regular fueling is not possible,
such as satellites, navigation buoys, heart pacemakers. Californium
is used in radio therapy. The transuranic nuclei decay mostly by
emitting α particles (helium nuclei composed of two protons and
two neutrons), but the secondary emission of β particles (negative
or positive electrons) and γ radiation (nuclear x-rays) is also possible.[3.10] Beta and γ emission, on the other hand, are the predominant decaying modes of the fission products.

Precious The quantitative distribution of the various waste products for a
waste pressurized water reactor (PWR) are shown in Table 3.2 and in
Fig. 3.9. In particular, 170 kg of fissile plutonium are contained in
the waste that is produced annually by one reactor of this type. If

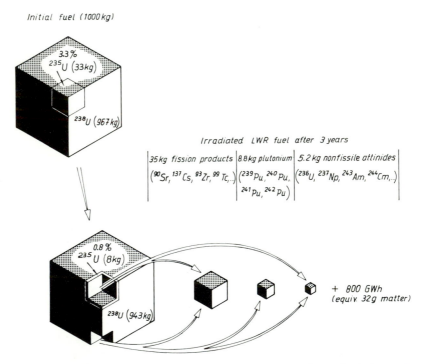

Figure 3.9 Transformation of PWR nuclear fuel at the end of the cycle,
before removal from the reactor. (Adapted from Ref. 3.5.)

extracted, this plutonium represents precious nuclear fuel.

When the maximum burnup is reached (i.e., after about 3 years for a PWR), the fuel rods containing the radioactive waste and the unburned uranium oxide are deposited in special water tanks within the power station for several years to allow abatement of the radioactivity and the heat generated thereby.

Thereafter there are various solutions. The most straightforward one is to dispose of the spent fuel rods directly for final storage over more than 100 000 years. As 95% of the initial uranium is still in the rods (Fig. 3.9), this solution is the most wasteful with regard to uranium resources. Alternatively, the rods can be sent to a reprocessing plant where the uranium and plutonium will be recovered, *No conservation* and the highly active waste is extracted for minimum volume final storage. In between these two extreme options, there are other intermediate solution, for example, to temporarily store the used fuel for some tens of years before eventual reprocessing, or before final storage. Which option to choose is the subject of much discussion and controversy, as it depends on such arguments as the present cost and availability of uranium fuel, the amount of exploitable uranium resources, final storage problems, or the future role of nuclear energy altogether. It would seem, however, that at a certain point reprocessing of spent fuel is a necessary condition to avoid squandering of one of our planet's major energy resources, and ultimately for nuclear energy to become one of the relevant *Squandering* energy sources of the future. *practice*

Various chemical extraction processes have been developed and applied since the first one was put into operation in a pilot plant in Oak Ridge, Tennessee in 1943. They were developed to extract weapons grade plutonium from metallic fuels at low burnups (700 to 3 000 MW·day/t); later the extraction of the unburned uranium for reuse in new fuel became an important goal, too, since the uranium quantities stored in the partially treated solutions were much larger than what could be mined at that time.

At present, the PUREX process (plutonium uranium refine by extraction) is by far the most important and is actually the only one considered for civilian application.[3.11] Originally, it was developed for metallic fuels of weapons relevance, but it was later adapted to *Refined* the oxide fuels used in civilian reactors. The application of the *ashes* PUREX process to civilian spent fuel is more difficult than to weapons grade fuel, since the former is many times more radioactive, a fact which already causes difficulties with chemical reaction dynamics. As far as exclusively civilian plants for reprocessing oxide fuels (UO_2) are concerned, there is at present (1982) only one in operation: It is located in La Hague (France) and has a nominal capacity of 800 tons/year, although various difficulties have limited the total amount reprocessed up to now to about 400 tons only. However, the design, and in some cases the construction of new plants for civilian fuel is under way, albeit at a slow pace. In fact, there is at present hardly any commercial justification for building reprocessing plants to extract plutonium fuel from the spent nuclear wastes.

The spent radioactive fuel rods from power plants are being stored in large water pools, mostly at the reactor sites, waiting for reprocessing to get started. About 7000 tons of spent civilian fuel are stored in the United States alone. In comparison, one must also mention the approximately 300 000 m^3 of liquid radioactive waste from the partially reprocessed fuel derived from weapons grade *Wasteful* plutonium production that are stored in large tanks mostly at the *delay* Hanford Reservation in Richland, Washington.[3.12]

The delay in reprocessing civilian fuels is due to a variety of economic, technical, and political reasons[3.13] and also to continuously varying and tightening safety standards.

The extracted highly radioactive waste with long decay times is sent for final storage, typically over 500 years for the fission products and 100 000 years for the nonfissile actinides. For this purpose the waste is, for example, vitrified and deposited in places selected for their hydrographic and geological stability. Extensive studies are under way to determine the most convenient methods and places for storage, so that the radioactive products during their long lifetime do not reach the surface and enter the biosphere.

A large nuclear power plant produces an average of 1.2 tons year of high activity waste (Table 3.2), which in practical terms represents a storage volume of about 3 m^3. If half the electrical energy produced in the European Community (a total of 1.1×10^{12} kWhe in 1980) were generated by nuclear power stations, the high-activity waste would amount to less than 0.4 g per inhabitant and year.

On the other hand, the extracted fissile plutonium, which is among the most troublesome radioactive components in the waste because of its 24 000 years of decay time, is reused as new reactor fuel and thus eliminated. Nevertheless, in the reprocessing procedure and in the subsequent plutonium fuel fabrication, medium-activity waste of about 100 m^3 per 1 GWe power station and year is produced, which, because of its 2 to 5% plutonium content, must also be considered for long-term storage. For this reason, the very long *Final* term final storage space required by the reprocessed fuel cycle of a *storage* 1-GWe power plant per year is probably not very much smaller than the one necessary to store the 28 tons of unprocessed fuel rods.[3.14]

For completeness' sake (Figs 3.4 and 3.5), one must mention an additional amount of some 300 to 500 m^3/year of very low- to medium-activity waste, with relatively short decay times, which are produced in any case by the operation of the nuclear reactor (such as cooling water filters, clothes, substituted mechanical components). This material is treated in different ways, being generally stored locally (for example, intermixed with concrete) or dumped in the oceans (for low-activity items). At the end of the reactor lifetime, after 30 to 40 years, the active components located near the reactor core must also be disposed of.

In a quite different perspective, the complete fuel cycle of nuclear fission energy can be seen as a process whereby the natural radioactive uranium and thorium elements with very long lifetimes are collected and then transformed by fission and neutron capture into a somewhat larger number of other radioactive nuclei with substantially shorter lifetimes (Table 3.3). This means that the radioactivity *Different* (i.e., nuclear decays per time unit) of the waste products is for a *perspective* certain time interval larger than that of the corresponding mass of uranium from which it is derived. This time interval amounts to several million years; seen in this time scale, then, the burning of uranium effectively corresponds to the elimination of radioactive products from earth.[3.5] In any case, the radioactive waste material can in principle be stored and kept under control, while the radioactive uranium and thorium minerals, from which the waste is derived, are located over many areas of the earth's crust.

3.3 SAFETY ASPECTS

Nuclear safety is an issue where very positive experience interrelates with statistical projections, opinions, and concern. Safety can be related to three distinct problems:

- Major reactor accident

- More or less regular release into the biosphere of minimal quantities of radioactive elements from reactor operation and its related fuel cycle
- Nuclear weapons proliferation

Myth and reality

Let us first discuss in simple terms the adverse health effects of radiation, since this sets the level on which the whole safety issue *Decaying* depends. A radioactive product is characterized by its activity and *activity* its half life or decay time (see Table 3.3). The former involves the number of disintegrations with emission of particles and/or electromagnetic radiation per unit of time, measured in curies (37 billion disintegrations per second), or in the new becquerel unit (1 Bq = 1 disint./s). The decay time is the average time required for radioactivity to be reduced by half. The harmfulness of radioactivity to man depends not only on its level, but also on the type and the energy of the emission. Fission products generally give off electrons (β radiation) and electromagnetic (γ) radiation, while actinides decay by emitting α particles. The biological effects of α particles are more serious than those of β or γ radiation, but the α act only if inhaled or ingested because of their limited penetration capability *Beware* (see note in Table 3.3). Radiation exposure is measured in units of *the rem* rem, or by the new sievert unit (1 Sv = 100 rem).

A one-time short exposure to 500 rem of γ radiation, e.g., is fatal to approximately half of the people so exposed. The problems with reactor safety, however, are the health effects (principally latent cancers and unfavorable genetic effects) caused by much smaller doses on a large number of people over a long period of time. Here the problem becomes controversial, particularly with respect to the so-called linearity hypothesis. According to the latter, the adverse effects are proportional only to the radiation dose absorbed, independently of the number of people exposed, the length and intensity of exposure, or any other factor. The merit of this hy- *Easy* pothesis is that it is easy to estimate the effects of radiation. In *hypothesis* contraposition, one could expect that the permanent damages become insignificant below a specific radiation threshold, which is related to each individual as a consequence of a healing effect in time. But, where is this threshold situated?

Table 3.4   Radiation doses per year

		Milli-rem/year
*Typical average individual doses**		
From natural radiation exposure		250
Cosmic radiation	35	
Terrestrial sources (from soil, building materials, etc.)	65	
Radon (inhalation)	120	
Body internal (ingested through food)	30	
From artificial radiation exposure	155	
Medical irradiation	150	
Fallout (from weapon tests)	4	
Nuclear energy (production, research)	<1	
Other sources	<1.3	
Admissible doses		
Population		
Maximum per unit of population (with a maximum of 5000 milli-rem in 30 years)		500
Exposed workers		
Maximum for occasionally exposed workers		1500
Annual average (with a maximum of 3000 milli-rem per quarter)		5000

*Values are for Switzerland, 1982. Single values differ widely, according to world areas, standard of living, social and occupational status. (1 sievert = 100000 mrem)

At present there are no clearcut scientific answers to these questions, precisely because the effects are extremely limited, and therefore difficult to quantify and assess correctly. Nevertheless, there is a certain consensus among experts that the linearity hypothesis represents a gross upper limit on the adverse effects caused by low doses of radiation, and that latent cancers are the most significant of these, with the incidence of one latent fatal cancer per 10000 man-rem of exposure.

Exposure to radiation is, on the other hand, a fact of life and has always accompanied—even substantially fostered—the evolution of life on earth. In fact, the sun and the stars emit radiation, to which *A fact* man is exposed. Moreover, radioactive materials are found on our *of life* planet, which can be ingested by man or which can irradiate man externally. In addition to these natural sources of radiation, man undergoes an additional dose of the same order for reasons of preventive and remedial medicine (Table 3.4). In relation to this, the nuclear industry contributes on average much less than 1% to the total average exposure of about 0.2 rem. It can be estimated that under the assumption of the pessimistic linearity hypothesis and of cancer risks as known today, present nuclear power cuts less than 15 min off the typical person's life span. Even if this average figure related to normal reactor operation were 100 times worse, it should hardly be the cause of major alarm.

Whatever the case may be, the effects of radiation should be viewed in the broader perspective of the damage inflicted by man on the biosphere, from the use of energy sources other than nuclear, and also from the manufacture and proliferation of chemical substances, not the least those from the pharmaceutical industry.
Better *life* *through* *chemistry?* Chemicals can, as is well known, cause mutations in the somatic cells of sexually reproducing organisms; much human cancer is probably caused in such a way. But can some chemical substances also cause inheritable mutations? This is a topic of present research in this field.[3.15]

There are also sensitive safety aspects at various phases and levels of nuclear weapons programs: during nuclear explosive fabrication, at the testing stage, and during storage. With regard to nuclear weapons-related testing, it can be estimated, for example, that by 1982 about 6 tons of plutonium had been dispersed in the atmos-
Weapons *testing* phere from the 440 nuclear detonations there, and that somewhat more than that quantity will have been deposited in the earth's mantle by an additional 880 underground detonations.

Perception of risk, however, is a more personalized and localized matter, which particularly relates to the fear of catastrophic reactor accident. It is thus necessary to give a more quantitative assessment of the probability of such accidents.

Generally speaking, evaluating the probability (which is never zero) of a serious accident can be carried out statistically, i.e., on the basis of accidents which have actually occurred in the past. For example, the probability of dying in a road accident in Switzerland,
Never *zero* as a typical European example, was one fatality per nearly 6000 inhabitants in 1977. A calculation of the risk associated with the exploitation of nuclear energy can today be based only partly on

experience. If such a calculation were made, the results would be very favorable in relation to other sources of energy. As of now (1985), that is, no member of the public has yet been either seriously injured or killed by a commercial reactor accident after more than 3500 reactor·year of operation (including the most serious mishap, which occurred at one of the reactors at Three Mile Island in 1979).

A thorough safety analysis of nuclear energy must therefore be based on a probabilistic (theoretical) risk assessment[3.19] rather than on experience, that is, by (1) identifying the real risk sources; (2) forecasting quantitatively the probability of malfunction of each individual component within these sources; and (3) establishing the (sequential) consequences of the malfunctions. Finally, acceptability criteria for these risks must be defined, as this is a necessary input to proceed toward a comparative assessment of nuclear energy versus other energy sources. Probabilistic risk assessment was applied in the Rasmussen Report commissioned by the American Nuclear *Quantifying* Regulatory Commission in 1974. On the basis of this technical *risks* approach, the report concluded that nuclear installations are very safe (e.g., the average risk of any American citizen dying as a result of an accident in any 1 of 100 nuclear reactors installed in the United States was estimated at lower by a factor of 10 000 than the risk of being involved in a car accident). Although today the commission no longer accepts its conclusions, the Rasmussen report constituted a pioneering first attempt to quantify the nuclear risk.

In future, every incident will have to be thoroughly analyzed and the results compared with statistical extrapolations (as has been done with the Three Mile Island accident) in order to gradually build up a valid assessment of the real risk involved. This process *Objective* will make it possible to carry out an objective comparison with the *comparison* risk of other energy sources and, in the final analysis, with the consequences of energy generation and consumption on a national or international scale.

Weapons-proliferation problem

Reactor operation together with reprocessing and enrichment technologies produce and separate fissile isotopes, which in principle can be used for either civilian or weapons programs.[3.17] One of the much debated energy-related problems consequently centers on the

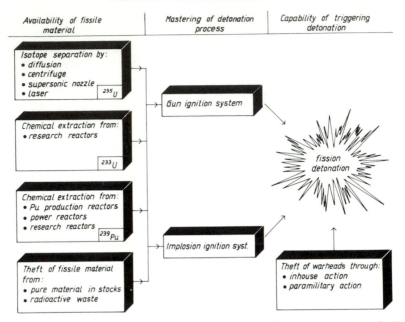

Figure 3.10   Possible means to reach the capability of detonating fissile materials.

concern that the development and diffusion of nuclear fission energy could implicitly contribute to proliferation of nuclear weapons, thereby increasing the risk of destabilization and, ultimately, of nuclear war. How far is this concern justified, and what can be done to control proliferation?

Crucial   A necessary condition to attain the capability of triggering a
condition   nuclear detonation (Fig. 3.10) implies:

● The availability of the appropriate fissile materials
● The mastering of the nuclear detonation process

In general, the availability of fissile materials is considered to be the crucial condition for proliferation. Once these materials are at hand in quantities and qualities as specified in Table 3.5, the ignition and detonation of fissile materials are considered within the grasp of a group of determined people with competence in physics and some technical skill relative to chemical explosives and electronics.

The situation is quite different when thermonuclear fusion weapons are considered. These are not included in the proliferation

Table 3.5   Critical masses of some fissile elements*

Metallic compound	Concentration %	Critical mass (kg of fissile elements) Bare	With reflector[†]
Uranium 235	100	47	15
(18.8 g/cm^3)	60 (+40 ^{238}U)	—	22
Uranium 233	100	16	5.8
(18.8 g/cm^3)			
Plutonium 239	100	10	4.4
(α phase,	70[‡]	13	4.6
19.8 g/cm^3)			

*From various sources, mainly *Rev. Mod. Phys.*, vol. 50, part II (January 1978).
[†] Reflector is typically 10 cm of uranium 238; the critical masses can be further reduced by a factor of up to 3 when strongly compressed (see text).
[‡] As extracted from reactor waste after a burnup of 30 GW$_{th}$·day, including about 30% plutonium 240.

problem because at present their ignition process is quite sophisti- *Fusion* cated and requires—in the first place—a fission detonation. The *weapons* fusion material is lithium deuteride (LiD, enriched with the lithium 6 isotope; see Chap. 5) and includes about 100 g of tritium to boost ignition. The availability of these materials, particularly the latter, poses quite different problems than those for plutonium, but on roughly the same factual difficulty level. Over the long term, ignition of a thermonuclear fusion detonation without a fission trigger, albeit with sophisticated technical means, cannot be fully excluded, al- though tritium would more than ever by required as an ignition booster.

All three major fissile materials considered so far (uranium 233, uranium 235, plutonium 239[3.1]) can be used as nuclear explosives. Technologies that allow the preparation of these materials are con- sidered "sensitive" with reference to the proliferation risks, i.e., isotope separation for uranium 235 and reprocessing of spent fuel for uranium 233 and plutonium 239. Heavy water production is also *Sensitive* considered a somewhat sensitive technology because a good moder- *technologies* ator is required to sustain a chain reaction in natural uranium. This could be either very pure graphite or heavy water. If the latter is available, plutonium can be bred from natural uranium, which can easily be produced and obtained at national levels, in contrast to enriched uranium, which is, at present, difficult to procure covertly. There is thus a potential equivalence between a heavy water production plant and an uranium 235 enrichment plant: Both can contribute crucially to the operation of nuclear reactors and thus to the breeding of fissile materials, where the latter could also be used as nuclear explosives.

The specific merits of fissile materials are deduced from three major arguments, which characterize a nuclear explosive:

- Fission physics
- Ignition process
- Production methods

The critical mass is a useful concept, which implicitly encompasses many relevant aspects of the fission physics. It represents the smallest mass of fissile material that can detonate, i.e., in which a chain reaction with fast, prompt neutrons can just be sustained.

Critical mass

The application of a reflecting blanket can reduce the critical mass by a factor of nearly 3, as indicated in Table 3.5 (the blanket reflects part of the otherwise escaping neutrons back into the reacting mass). Strong compression as obtained in sophisticated implosion devices can further reduce the critical mass; this is particularly true for plutonium. In conclusion, the critical mass can be as low as 2 to 3 kg for plutonium 239 and 8 to 15 kg for uranium 235; this corresponds, for example, to a plutonium sphere of 6-cm diameter at normal density.

Countdown to ignition

Ignition of a nuclear explosive consists in rapidly assembling two or more subcritical components of fissile material into an overcritical mass, and then injecting a burst of neutrons to start the chain reaction efficiently. In order to ignite uranium 235, the relatively simple gun method is usually applied to assemble the overcritical mass; whereas for plutonium the more elaborate implosion method is used because it implicitly allows a reduction of the critical mass by compression, and also because—being faster—it can cope better with the neutron background effect. The spontaneous fission of plutonium 240, which is inevitably intermixed with plutonium 239, produces constantly about 1000 neutrons per second and gram of this isotope. Plutonium 240 is produced by about one neutron capture out of four in plutonium 239. These neutrons can start a (near) chain reaction before the critical mass is fully assembled, thereby eventually degenerating the detonation process to an inefficient blast-off.

The production of weapons-grade uranium enriched to more than about 93% with uranium 235 can be achieved with most of the known isotope separation methods. Lower enrichment still allows *Skilled* detonation, but the necessary mass becomes rapidly intractable: at *hobbyists* 60% enrichment, the overall critical mass amounts to nearly 40 kg (Table 3.5). Mentioning the centrifuge separation method seems particularly appropriate, since within two or three decades the necessary technology could well become accessible at the level of skilled hobbyists.

Fissile plutonium is produced by transmutation in any uranium-based reactor (Table 3.2). However, the illegal diversion and handling of plutonium from spent reactor fuel or from any reprocessing *Denatured* stage thereafter is rendered difficult by the prohibitive radiation *explosive* level of the accompanying waste products, even if only as impurities (this difficulty can be enhanced by artificially contaminating or denaturing the plutonium). There is also the isotope problem. Weapons-grade plutonium contains less than about 7% of plutonium 240, and supergrade plutonium for sophisticated (small) weapons less than 2%; this requires a burnup of the fertile fuel in the reactor of less than 3 or 0.7 GW·day/t, respectively.

The efficient use of nuclear fuel for power production implies, on the other hand, a burnup that in light water reactors is typically 30 GW·day/t (Fig. 3.9), in which case the plutonium 240 content amounts to 25%. The question is: Can plutonium produced in *Bulky* civilian power reactors with a typical plutonium 240 content of 10 to *implosion* 30% and extracted in reprocessing plants be used as nuclear explosive? The answer is: Yes, it can be detonated, albeit inefficiently, and with bulky, impracticable implosion systems. A possible alternative is to extract and eliminate the plutonium 240 by isotope separation (for example, by laser action), thereby obtaining weapons-grade plutonium from normal civilian reactor waste.

Covert procurement of plutonium cannot be excluded within normal and even safeguarded reactor or fuel-cycle operations, for example, by using special plutonium-producing fuel rods with low burnup, or by the just-mentioned plutonium isotope separation method, or by straightforward stealing from weapons-related factories.

Confusing opinions On these and some additional facts rest the main technical aspects of the proliferation problem, which entail a vast and, as far as the public is concerned, often confusing spectrum of opinions by specialists. In some cases these worries result in the propensity for renouncing nuclear energy altogether.

From the arguments presented one may conclude that the proliferation problem cannot be avoided by technical means alone; this will remain so, irrespective of whether the civilian nuclear program, including reprocessing, is expanded or stopped completely. A determined group of people at national level can find solutions to evade all inspection procedures or technical fixes devised up to now, particularly if it is not worried by the risk of being discovered at a *Technical* certain phase. Nevertheless, more stringent safeguard systems, res- *fixes* trictions in applying nuclear power, and avoidance of reprocessing could undoubtedly reduce the proliferation level, but only for so-called terrorist groups, and on the short term.

Table 3.6   Some information on the worldwide nuclear proliferation potential

- Nations that have detonated nuclear explosives:

  United States, Soviet Union, Britain, France, China, India

- Nations that most likely possess detonable nuclear devices, without having detonated them yet:

  Israel, South Africa

- Some nations that have the knowhow and the industrial potential to build nuclear bombs within 2 to 3 years, if they wanted:

  Canada, Sweden, Italy, Japan, Germany (FRG), Switzerland, Belgium, The Netherlands

- Some nations that show interest in the processes required for preparing (eventually detonating) fissile materials:

  Pakistan, Argentina, Taiwan, Brazil, Libya, Iraq, Iran

- Some nations that have nuclear capability but refuse to sign the Non-proliferation Treaty[3.18]:

  Argentina, Brazil, Chile, China, France, India, Israel, Pakistan, South Africa

If blackmail by "terrorist" groups with effective or only fictitious nuclear devices cannot be excluded in future years, it is likely that the real risk of triggering events that could lead to nuclear war will come from small, unstable, or arrogant national entities, and last, but not least, from the superpowers, whose "legal" behavior is not always free from proliferating attitudes. In addition, the management of tens of thousands of critical nuclear assemblies ready to be fired (it is estimated that the United States alone possess about 30 000 of them deployed or on stock) leaves margins open for mishap, even under the most stringent safety measures.

Terrorists and diplomats

In conclusion, then, one can say that the avoidance of nuclear proliferation, and more specifically of the risks of triggering a nuclear war, must basically be attained by institutional means, with little room for technical fixes (Table 3.6). As a positive move in this direction, one can mention the Non-proliferation Treaty, signed at the beginning of 1985 by 130 states.[3.18] However, even this treaty is at present somewhat languishing and requires a fresh push to become really significant.

CHAPTER 4

Energy Tomorrow

The planet earth traveling through space is richly endowed with raw materials and energy resources. New forms of energy—or older ones with much improved efficiency—are available or ripe for exploitation. In order to present significantly the broad spectrum of available options, the contents of this chapter must per se be wide-ranging. Consequently, it includes solutions which depend, on the one hand, still on the outcome of physics research, and on the other, on simple technologies (such as that of the heat pump) which have been in the public domain for a long time and are awaiting acceptance of their effective economic viability. What counts in the medium term are mainly economic considerations of the new energy technologies, and these are generally approached along a tortuous path of trial and error.

4.1 SPACESHIP EARTH

Planet earth can be seen as a spaceship where the inhabitants have to contend with the same problems of survival as present-day astronauts. They must use the stocks on board and those obtainable in space, not to mention the problem of waste processing and the inherent psychological strain of living in confined areas. The stocks on board our earth are the raw materials, the most important of which is energy.

Energy bank   As discussed in Chap. 2, the source of nuclear energy—the nuclei of uranium and thorium for fission, and of deuterium and lithium for fusion—were put into the earth at its formation some 5 billion years ago, while fossil fuel deposits are generally only some hundred million years old, being the decomposition product of organic matter.

Direct solar energy is particularly interesting, since it is picked up continuously as the earth travels through space; it is, by definition,

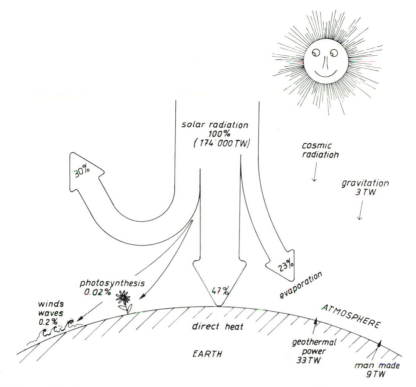

Figure 4.1 Power balance on spaceship earth's surface;[4.10] as the earth's temperature remains about constant, the overall power influx is offset by a reemission of the same amount of power in the form of long wave (infrared) radiation (1 TW; 1 terawatt = 10^{12} W).

renewable (Fig. 4.1; see also Fig. 2.4). In addition to direct radiation, solar energy is also used in significant secondary forms, such as through photosynthesis (from which fossil fuels are derived), evaporation (on which the exploitation of hydropower is based), and atmospheric heating (which is responsible for wind generation). *Oil and flowers*

The solar power reaching the earth's surface is 10 000 times greater than the average power currently generated by man.[4.10] In overall terms, therefore, the world's consumption of energy is still comparatively moderate. However, as we have seen, the safety and environmental problems on earth derive from generating and transforming energy into forms which can be used by man, and, in

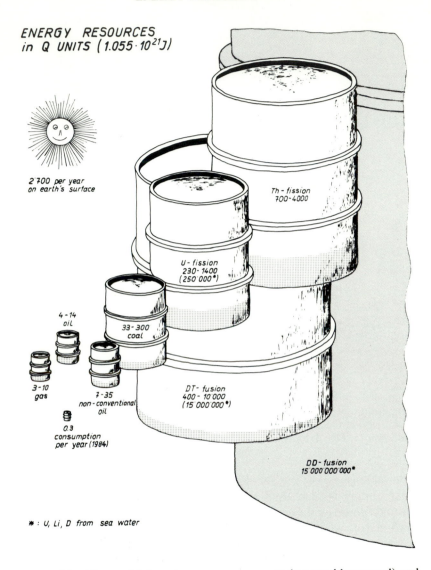

*ENERGY RESOURCES
in Q UNITS ($1.055 \cdot 10^{21} J$)*

2'700 per year
on earth's surface

Th - fission
700 - 4000

U - fission
230 - 1400
(250'000*)

4 - 14
oil

33 - 300
coal

3 - 10
gas

7 - 35
non - conventional
oil

DT - fusion
400 - 10'000
(15'000'000*)

0.3
consumption
per year (1984)

DD - fusion
15'000'000'000*

✳ : U, Li, D from sea water

Figure 4.2   The earth's long-term energy reserves (reasonably assured) and resources (extrapolated and hypothetical). The barrel's volume is proportional to the latter, the larger number in each pair of figures, which includes also the former (indicated as the barrel's shaded bottom). The resources in the sea, indicated by an asterisk, are too large to illustrate. (See text and Refs 4.1 to 4.6 for more details).

particular, from the local concentration of energy production.

An evaluation of the overall fossil and nuclear energy resources on earth is difficult, since it depends on various degrees of estimates and speculations regarding both extraction costs and the amount of identified, or as yet undiscovered, ore deposits. It is customary to adopt the so-called McKelvey classification chart, where geological knowledge is labeled as "identified" (proven, probable, and poss- *Resources* ible) and "undiscovered" (hypothetical and speculative); whereas *and* the economic interest is divided into "economically recoverable" *speculations* and "subeconomic" categories (at various levels).

Placing the emphasis on long-range energy availability, we adopt here a much simpler scheme that is expressed in Fig. 4.2 by a pair of figures. The lower figure refers to the energy content of reserves, which have been fairly identified (including in the upper limit some inferred reserves) and with extraction costs estimated within a factor of about 2 of present ones. The higher figure includes hypothetical (or in the limit even some speculative) resources with costs estimated within a factor of about 2 to 3 of present ones, and also includes the reserves just mentioned. A third figure marked by an asterisk indicates resources contained in the oceans[4.24]; whereas the extraction of deuterium is now competitively possible, and that of lithium may become so over the long term, uranium extraction from seawater— *Water* due to its extremely low content—remains a speculative under- *as* taking (see Refs 4.1 to 4.6 for further details). *fuel*

Finally, to obtain the amounts of useful energy, the mentioned figures are reduced by taking into account mining, refining, and burn efficiencies. It will be noted that we use the term "reserves" for the conservative estimates, whereas "resources" refer to the more hypothetical ones. The overall picture is, of course, crude, but it may be helpful in providing a significant overview on the earth's energy reserves and resources. Other studies arrive at somewhat different figures,[4.11] but the overall situation remains confirmed.

4.2   THE QUEST FOR NEW ENERGY SOURCES

Future scenarios must consider both the substitution of today's widely used fuels, such as oil, with new energy sources, and the overall increase in energy consumption. This requires

- The availability of new energy sources
- The means of exploiting them competitively

Full   As we saw in Sec. 4.1, energy resources on earth are more than
power plentiful. The real problem concerns the exploitation processes.
ahead Their merits can be evaluated with regard to commercial competitiveness, a criterion which becomes somewhat vague when
applied to a process lying 20 or 40 years in the future. A more valid
criterion is a positive energy output over the lifetime of the process,
meaning that the net energy delivered to the users must be larger
than the energy required to explore and prepare the energy source,
including the plant structure and materials necessary for its exploitation, and the related transportation needs.

As we shall see, most of the so-called new sources have been
known and in some cases exploited for decades or centuries. Present
New research and development programs are actually devoted to first
and trying to fullfill the necessary energy balance, e.g., by increasing the
old cycle efficiency or reducing the amount of the materials used, and
then to devise paths of development which will gradually establish
their commercial competitiveness. This is particularly true for sources
with low energy or power densities, such as solar radiation.

All industrialized countries are channeling considerable effort into
energy research and into the development and demonstration of the
necessary technologies. The development of new sources is of particular significance in Europe, which is comparatively poor in energy
reserves.[4.30] For instance, Europe has only about one-quarter as
much coal as the United States and only one-tenth as much oil as
the Middle East.

In search   Major research, development, and demonstration activities on
of new new energy sources or transformation processes can be characterized
sources by four lines of action that will be discussed in following sections:

- Energy conservation, both at source and at consumption
- Advanced nuclear fission reactors
- Solar energy
- Controlled thermonuclear fusion

The potential relevance of new or alternative energy sources also
depends on the energy density involved in the extraction or transformation process (Table 4.1), which contributes to determine the
technology level required for their exploitation, their commercial

Table 4.1 Equivalences of some energy sources*

Energy equivalence of 100 000 kWh or 3.6×10^{11} J		Total energy content, J/ton
12 t	coal	29.3×10^9
8 t	oil	45.4×10^9
4.5 g	uranium	82.2×10^{15}
1 g	deuterium	350×10^{15}
4 mg	matter	90×10^{18}
25 000 m²	1 day solar radiation on earth's surface	—

*9 000 liters (8 t) of oil are required to heat four apartments in Central Europe during the winter, or to drive a car twice around the world at the equator. References are made to total, potential energy content; useful energy is generally lower, since it is determined by energy transformation and burn efficiencies. Solar energy: Based on an average radiation of 0.17 kW/m². Fusion: Based on 43.3 MeV from the fusion of 6 deuterons; note that the deuterium contained in 1 liter of seawater is equivalent to over 300 liters of oil. Fission: Based on 200 MeV energy per fission. Matter: According to the relation $E = mc^2$, the total world energy requirement in 1984 is equivalent to more than 4 t of matter.

competitiveness, and even their level of risk (Sec. 1.4). Nuclear energy density is roughly 1 million times larger than that of chemical (combustion) energy, whilst complete annihilation of matter, according to Einstein's energy-matter relation, further augments energy density by a factor of less than 1 000; this is in direct proportion to the strength of the forces, which determine the chemical bonds (the outermost electrons in an atom), the bonds of neutrons and protons *Universal* in a nucleus, and the cohesion of the nuclear matter itself. *forces*

Technical development and demonstration work is also devoted to many other sources of energy; in some cases this work refers to new exploitation forms, in others to efficiency increases of sources which in part have already been successfully exploited,[4.15] such as:

- Coal
- Nonconventional oil
- Wind
- Tidal energy
- Geothermal energy

It is customary to distinguish the so-called renewable sources, which include the last three sources plus solar energy. This is not a

Renewable resources   satisfactory characterization, since nuclear fusion and fission energy are also renewable in the sense that they will in practice last for ever.[4.1,4.2]

The production of synthetic fuels ("synfuels") from coal and from oil shale rose in the seventies to much publicized pilot projects supported by some of the major oil companies. However, interest in these undertakings has now practically vanished, because present *Publicized* trends of energy cost evolution show that competitiveness of syn-*synfuels* fuels will not occur in the near future. Meanwhile, technical improvements concerning extraction of solid fossil fuels regard, for example, the underground exploitation of coal through in situ gasification,[4.17] or its transport through pipelines, the coal being transformed into slurries.[1.5] Fuel extraction from thick oil, tar sand, or oil shale[4.4] remains an important step for tapping the enormous energy reserves which are contained therein.

Wind power dates back thousands of years and is available on a massive scale, yet it is difficult to exploit on account of its low power *Exploited* density. It is estimated that the average power available from shifting *Aeolus* air masses all over the planet is 1.8×10^{15} W, i.e., approximately 1% of the solar power (which generates the winds) absorbed by the earth. This corresponds to 3.5 W/m^2 of surface. However, wind power is concentrated in some exposed locations where average power densities of 500 W/m^2 of vertical surface at a nominal height of 25 m from the ground are not uncommon.

Currently, the most advanced technology is represented by the large, horizontal-axis wind turbine with two rotor blades.[4.20] For example, the now operating generation of wind turbines developed by NASA has a two-bladed rotor spanning 90 m, which starts to generate power in a 23 km/h wind and achieves power ratings of 2.5 MW in a strong breeze (45 km/h or force 6 wind) (Fig. 4.3). Practical exploitation of wind power in the near future should be limited to serving remote, windy areas with small-to-medium size turbines.

Geothermal heat is an other interesting energy source, since, in principle, it can be tapped nearly everywhere on earth.[4.19] However, like the winds or solar power, its drawback is the low-power *A hot* density generally available, making it very difficult to exploit com-*planet* petitively. From the earth's core of hot magma, heat flows by conduction (and exceptionally by convection) to the surface, thereby establishing a mean temperature gradient of 2.75°C per 100 m depth. The mean geothermal power that flows through the earth's

Figure 4.3   Large two-bladed wind turbine on a 60-m-high tower; rotating at 17.5 revolutions per minute, it provides an electrical power output of 2.5 million W. (Courtesy: U.S. Department of Energy.)

surface is 0.063 W/m^2, a tiny fraction (1:2 700) of the mean solar power flow that reaches the surface (Sec. 4.4). The ideal procedure is to drill holes into the crust to reach high-temperature zones and extract the heat by means of a fluid (e.g., water) and thus bring it into a thermal cycle at the surface. In an ideal situation, temperatures of 100 and 300°C are reached at depths of 3.3 and 10 km, where the water pressure, corresponding to a free column of this height, is about 320 and 980 atm (the adjoining rock pressure is higher by nearly a factor of 3, because the mean specific gravity of the crust material is 2.7 g/cm^3).

Tapping hell

Drilling techniques have improved appreciably in past years; drilling speeds in granite of several meters per hour are currently being attained. The costs are high: more than $15 million for a pair of 5-km-deep wells (1982). However, the major technical difficulty remains the heat exchange between the hot dry rock and the water, i.e., the necessity to largely augment the contact surface. When there are no natural fissures (which are rather rare), this must be done by artificial fracturing of the rock, for example, by means of hydraulic pressure or explosive action. The overall geothermal energy available in the earth's crust down to a depth of 10 km has been estimated at more than 1 million Q units (see in comparison Fig. 4.2). Unfortunately, as we have seen, the low density of most of this energy, the inherent problems of extracting heat from the hot compact rocks, and the unknown effects of possible seismic reactions make the practical application of this type of energy very difficult.

The *Geysers* The earth's crust is not always regular. In fact, in some places, the hot magma comes close to the surface—in some eruptions it actually breaks through—and thus heats overlying rocks to much higher temperatures. In some fortunate cases, natural water seams in contact with these rocks produce hot vapor which can be sealed off in impermeable underground reservoirs, or they can erupt freely into the atmosphere. These geological formations are the sources of presently exploited "natural" geothermal power. The installed capacity of geothermal origin was a global 2.8 GWe in 1982 with 0.44 GWe operated by the Italian Electricity Board at Larderello and Monte Amiata in Tuscany. Exploitation of natural geothermal power started in Larderello in 1904. Other electricity-generating plants are located at the Geysers (California), at Wairekay (New Zealand), and at Matzukawa (Japan).

4.3   ENERGY CONSERVATION

Real saving of energy is achieved when comparable effects and commodities are obtained by using up less primary energy. This can be achieved by means of:

- Reducing consumption, where possible
- Increasing the source and/or consumption efficiency[4.14]

Energy conservation depends upon a broad spectrum of options. Real conservation is attained as a result of systematic long-term policies and investments. With the exception of the straightforward reduction of consumption (e.g., using mass transportation instead of private cars, or lowering the heated room temperatures)[4.16], additional fuel economies can be attained only at the cost of substantial investments and of new equipment, which, in turn, require energy for their manufacture.

Energy conservation is a much discussed subject, as it is easy and appealing to talk about. It is often presented as a hidden but most *Easy* immediate new "energy source." Some success has been obtained in *to* past years, notably in areas where rather wasteful energy practices *talk* prevailed in the plentiful sixties. For example, in the United States, an efficiency increase in the use of energy of around 10% has been obtained in new car mileage, home heating and industrial output, even before any substantial investment in retrofit or development programs has had time to take effect.[4.7] Generally speaking, however, conservation is not at the level most people would like to have it; the cost of energy is still comparatively low in many economic contexts. In this regard, it is interesting to remember that the tax on fuels for private transportation exeeds the primary fuel's cost in most European countries. One could resume these aspects into one simple question: Why do people like so little to save money by saving energy?

The economic aspects of energy use and conservation are complex and intertwined. An increase in the primary energy cost clearly *Intertwined* augments the cost of materials that require energy for their pro- *materials* duction (see Table 1.3). Therefore, the cost of materials used in conservation levitates with the increase of fuel costs, and any specific conservation process may or may not become economically viable in these circumstances.

There are situations where conservation seems particularly appealing. The huge amounts of waste heat discharged into the environment by fossil fuel or nuclear electropower plants could be used as process heat in industry and for district heating (see Fig. 1.4). It is instructive to discuss this option in more detail.

Cogeneration

Most of the electricity produced in the world (with the exception of hydroelectricity) is generated in plants where heat from a fossil fuel *Steamy* burner, or from a nuclear reactor, is used to boil water and raise *electricity* steam to temperatures of 300 to 500°C at typical pressures of 70. This steam is then expanded through a turbine that drives an electrical generator (Fig. 4.4). To obtain the maximum conversion of

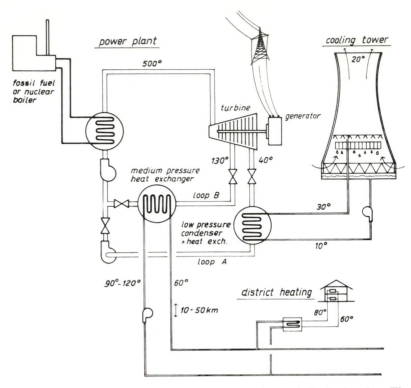

Figure 4.4 Combined electrical power generation and district heating. The figure shows a flexible plant that can generate electricity only (loop A) or operate on a cogeneration mode with steam extracted part way along the turbine (loop B).

thermal energy into mechanical and electrical energy, the steam is expanded to as low a temperature and pressure as possible. Thereafter it is condensed and the waste heat thus liberated is released at low temperature into a cooling fluid and finally into the environment.

Inevitable   That typically two-thirds of the primary energy used in such plants *loss* must be rejected as waste heat is a fundamental aspect of any process (operating in the mentioned temperature range) converting thermal energy into mechanical and eventually into electrical energy (Sec. 1.2). In general, the temperature of the waste heat (typically 40°C) is too low to be of any direct use, since even room heating applications require temperatures of 40 to 80°C; when industrial

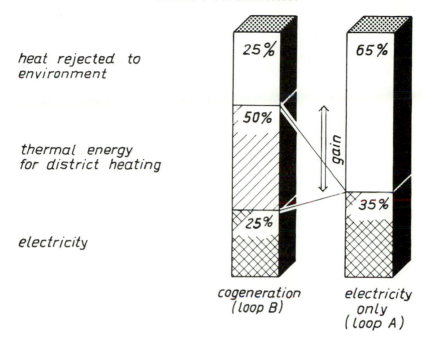

heat rejected to
environment

thermal energy
for district heating

electricity

25%

50%

25%

65%

gain

35%

cogeneration
(loop B)

electricity
only
(loop A)

Figure 4.5   Gain in the efficient use of energy in cogeneration power plants
versus electric-only power plants (compare with Fig. 4.4).

process heat is the purpose, then vapor at more than 150°C is
usually required.

Cogeneration refers to a procedure for exploiting the waste heat
and is thus one important possibility for increasing the efficiency in
energy consumption (Fig. 4.5). The cogeneration procedure is
particularly convenient when district heating is inserted into the
cycle, since this alone is an efficient way or providing the thermal
needs of a large community from one or more large centralized heat
sources. Combined power generation and room heating can also be *District*
applied at much smaller levels, where, for example, the mechanical *heating*
power generator is a small diesel engine and the heating energy
comes from its discharge heat.

District heating is an old and simple technology. It developed
rapidly from the end of the last century, particularly in the large
American cities. A well-known example is the New York City steam
district heating system dating back to 1880. As electrical power
plants, invented by Edison, commenced operation soon afterward,

and since domestic and industrial energy users in the big cities and in industries wanted both electricity and heat, cogeneration was applied and exploited a few years later.

There are various options regarding heat transport over tens of kilometers from the central power plant to the rooms to be heated.[4.22] Many studies prefer the closed water/vapor loop with initial temperatures of around 90 to 450°C. The thermal energy to this loop is delivered—through condensers and heat exchangers—by vapor extracted from the turbine prior to full decompression. In Fig. 4.4 the two basic operation modes of this type of power station are shown schematically and compared:

- Electricity production only (loop A). Electricity production amounts to typically 35% of the primary thermal energy consumption, the remaining 65% being lost to the environment.
- Combined electricity-heat production (loop B). Electricity production is reduced to typically 25%, but the remaining thermal energy is used with high efficiency for room heating (50%), and only 25% is lost to the environment (in the power plant and in the district heating network).

However, the energy conservation due to the increased efficiency (Fig. 4.5) of primary thermal energy consumption in loop B requires the following investment: Installation of additional electricity-generating capacity to compensate for the reduction of 35 − 25 = 10% (more than $2 million per MWe in a nuclear plant); additional *Gain* power plant equipment for heat extraction ($30 000 per MWth); *and* pipeline loop for heat transport over 10 to 50 km. The necessary *losses* district heating capillary distribution system represents a separate investment, since its particular merits are independent of electricity-heat cogeneration.

Heat derived from cogenerating nuclear power stations is already applied in many places. For example in Switzerland, the power station of Gösgen delivers 9 MW of process vapor at 210° C over a 1.8 km long pipeline to a cardboard factory, thus avoiding the combustion of about 20 000 t of oil per year; also, the nuclear power station of Beznau delivers heat through a distribution system (which cost $30 million) to a population of 15 000 inhabitants of the Aar valley, thus saving about 16 000 t of oil per year. Detailed plans exist to extend this system up to the city of Zurich, and a study suggests delivering heat to 1 million people living within about 20 km of the feeding nuclear power stations.[4.8]

Heat pump

The heat pump is a device that uses mechanical energy to transfer, against the natural tendency, thermal energy (heat) from a cool medium to a warmer one (Fig. 4.6). In this sense it works much like *Uphill* a pump driving water uphill.[4.25]   *flow*

The heat pump runs on a traditional evaporation-condensation cycle, a concept first proposed by Lord Kelvin (William Thomson, 1824−1907). In addition to the mechanically or electrically driven vapor compressor, there are other means of pumping heat between the evaporator and condenser, such as chemical absorption or thermoelectric methods. In any case, the pump's heat delivery mode of operation is, basically, refrigeration in reverse. Air conditioning, for example, uses an electrical compressor to remove heat and moisture from warm indoor air to the outdoors; conversely, a heat

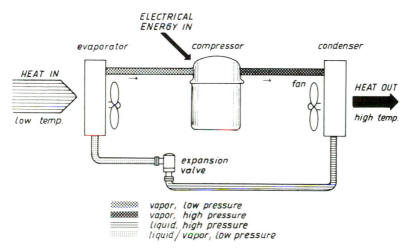

Figure 4.6 The heat pump is an important element for directing and concentrating heat flows to useful applications.

Low-grade heat from the surroundings (e.g., air or water) is absorbed by the vaporization of the circulating fluid (e.g., freon or ammonia). The vapor is then compressed to high pressure and to a suitably high temperature. The heat contained therein is transferred and delivered to its final destination as the vapor condenses to a liquid. The temperature and pressure of the latter are drastically lowered when the fluid passes through the expansion valve; the resulting liquid/vapor mixture is then ready to restart the cycle.

pump on heating mode uses energy to direct heat in the opposite direction.

The heat pump is of considerable interest in energy conservation because the amount of mechanical energy required for its operating is only a fraction of the total energy (heat) output. The performance ratio of available to required operating energy can have values of between 2 and 3 for small machines using outdoor air, and values of up to 6 are feasible with larger machines pumping through smaller temperature differences. The essential point is that the input energy— *Heat at* heat at low temperature—costs nothing since it is plentiful and *no cost* available everywhere, because any medium contains thermal energy in amounts that are about proportional to its absolute (Kelvin) temperature.

Operating a heat pump with a performance ratio of 3 or more, by using electricity produced with an energy efficiency in excess of one-third, results in an overall heating (or energy) efficiency of more than $3 \times 1/3 = 1$. This represents an extremely tempting scheme when compared with the mean efficiency of about 0.6 for a home oil burner for heating application. In spite of this advantage, heat pumps are not widely used yet. For as long as primary energy costs are as low as they are now, inefficient energy use is generally more convenient than the application of heat pumps with their higher capital investment and uncertain long-range maintenance costs.

4.4   SOLAR ENERGY

The sun is a nuclear reactor in which 600 million tons of hydrogen nuclei are fused each second into 595.8 million tons of helium; the missing 4.2 million tons of mass are transformed each second *Heavenly* into energy, which is emitted into space at the power level of *reactor* 3.8×10^{26} W in the predominant form of electromagnetic radiation, while a much lesser share is transported by neutrinos, charged particles, and magnetohydrodynamic waves. The characteristic of the radiation that reaches the outer atmosphere is as if it were emitted by an incandescent sphere with a radius of 700 000 km and a surface temperature of nearly 6 000 K: 47% of the energy is carried by waves in the visible spectrum (wavelengths between 0.4 to 0.76 μm), 8% in the ultraviolet spectrum, and 45% in the infrared spectrum.

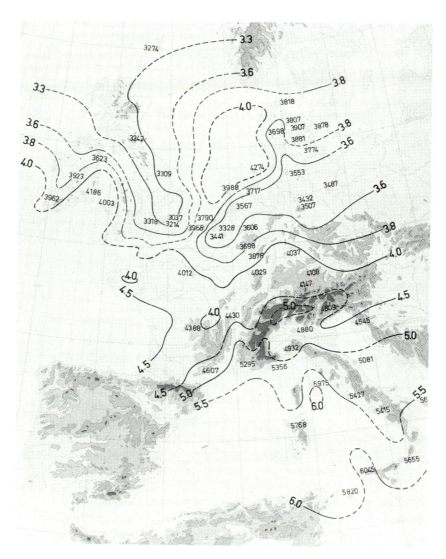

Figure 4.7 The mean solar energy in Europe, during the month of April, measured on a horizontal plane per day in kWh/m² (the values referred to specific places are in Wh/m²). (Courtesy: European Community, Brussels).

The overall radiation power density at the mean sun-earth distance of 150 million km is 1.35 kW/m^2 (the so-called solar constant), of which the time and seasonal average power density of 0.170 kW/m^2 of direct radiation reaches the earth's surface and peaks at roughly 1 kW/m^2 near noon on clear days.[4.10] In Rome the annual solar energy density average of direct radiation is 1 380 kWh/m^2; in other words, it is as if the energy equivalent of 14 cm of oil were to fall from the skies each year. Generally speaking, the further south one goes, the greater the solar energy at ground level, both because of a more favorable angle of incidence of the sun's rays and because of climatic factors (Fig. 4.7). The Swede John Ericsson in the United States between 1870 and 1884, besides developing a variety of thermal-solar engines, made the first accurate measurements of the solar constant.

The three main methods for exploiting solar energy can be classified as:[4.13]

- Photovoltaic conversion
- Photochemical processes, in particular photosynthesis
- Heliothermal processes

In addition to solar energy exploited through its direct radiation, hydro and wind energies are often also classified as solar. We shall not mention them here because their exploitation status is different, with hydro energy covering in 1984 about 5% of equivalent energy consumption (Sec. 1.3 and Fig. 1.7).

The exploitation of solar energy has to cope with two basic difficulties: the low power density and the irregularity of its supply. The former implies large area collectors, the construction of which re-
Soft quires intensive use of structural and radiation absorbing materials
hopes (in turn, this required energy). However, material consumption may eventually be reduced drastically by inventions or clever engineering, since what counts, i.e., what basically interacts with the solar radiation in the collectors, are thicknesses of only some micrometer of silver or alluminium in the mirror concentrators, or of a fraction of a millimeter for the semiconductor junction in the photovoltaic cells (as we shall see). Solar energy remains in principle a very interesting source, because it represents a clean energy available locally. Realistic comparison of the merits and disadvantages of solar energy shows that there remains, however, a thin dividing line between "soft," mythical hopes and hard reality.

The most immediate and convenient application of solar energy is
the decentralized, local scale production of low-temperature heat *Hard*
for heating and cooling purposes. Large-scale use of solar energy, *realities*
particularly for electricity production, is less immediate. To reduce
the related energy storage problems, the solar source will have to be
developed and incorporated gradually into energy systems comprising predominantly other sources of energy.

It is instructive to realize that a centralized solar power station,
generating an annual amount of electricity equivalent to that of a
1 GWe nuclear power station and located in the Mediterranean
(e.g., in Sicily), occupies a ground area of almost 100 km^2 (this
example is based on the assumption that the convertors have a 10%
efficiency rate and take up half of the useful area and, in addition, *An energy*
that the nuclear power station has a load factor of 0.75). The *island*
electricity thus produced yearly (6 500 GWh) would correspond to
approximately one-twentieth of the electricity now consumed in
Italy.

Development and demonstration of solar energy projects are
proceeding at present along many paths, of which the most important are mentioned with a slightly optimistic tilt in Table 4.2.

Photovoltaic conversion

Solar energy can produce electricity through conventional heliothermal cycles (e.g., power tower, see later) or through direct
photovoltaic or thermionic conversion (Fig. 4.8). In the latter, a
special cathode is heated by solar or nuclear energy, emitting electrons with sufficient energy to cross the space to the anode and
return through leads to the cathode; some heat is thereby converted
to electricity, at present less than 5%.

Of much greater practical importance is the photovoltaic conversion obtained by solar irradiation of some crystalline matter and
semiconductors, such as silicon, cadmium sulfate, or gallium arsenide. *Energetic*
The discovery of the photovoltaic effect in 1839 is credited to *wafers*
Antoine Becquerel. When light with appropriate characteristics
penetrates a silicon solar cell (a wafer of a thin layer of so-called
n-type silicon applied on top of *p*-type silicon), it liberates some
electrons from the crystalline bonds. These are driven by the electrical fields in the junction into the *n*-type layer and, through the

Table 4.2   Status of solar techniques

Technology	Main end product	Status
Photovoltaic		
Photocells	Electricity	Commercial, but high cost/no mass production yet
Solar satellite	Electricity	Uncertain application/ high-capital investment
Photochemical		
Photosynthesis (biomass)	Fuel	Ripe for exploitation
Direct conversion	Fuel/chemical products	New process and technologies required/application beyond the year 2000
Heliothermal		
High temperature		
Power tower	Electricity	Storage and cost problems/ applications beyond 2000
Energy farm	High-temperature heat	Partially ripe for exploitation
Low temperature		
Flat collector	Low temp. heat	Commercial collectors
Moderate concentrator	Cooling	Ripe for exploitation but cost problems
Indirect methods		
OTEC	Process heat/electricity	Low efficiency/cost problems/ appl. beyond 2000
Natural waters	Heat	Commercial

return lead (attached to the semiconductor by a collector), back into the *p*-type layer.

The direct generation of electricity by means of photocells is among the most interesting applications of solar energy.[4.21] Unfortunately, this method is very expensive, since the cells themselves are complicated to produce: A massive, much purified, cylindrical silicon crystal is cut into thin wafers which are transformed into *n*- and *p*-type sandwiches by special treatment, and to which electrical contacts are attached. The cost of these cells alone, in large lots, is still (1982) at $10 to $15 per watt of peak electrical power capacity.

Figure 4.8 The Solar Challenger has demonstrated impressively how much solar power can do by flying on July 7, 1981 on nothing but sunshine over 300 km from near Paris to the Manston Air Force Base in England.

The Challenger was designed and built by a group of American enthusiasts lead by Paul McCready, who had already succeeded a few years earlier with the first Channel crossing by human-powered flight. The 47-ft wing and the horizontal stabilizer were covered by 16 128 photovoltaic cells. (Courtesy: Gossamer Ventures, Simi Valley, California.)

It will require new cell types based on amorphous silicon deposited on sheets that are more appropriate for mass production. Unfortunately, these cells at present (1984) have an efficiency of only about 8% (18% for crystalline silicon); an interesting prospective would be to reach 15% for cells priced at below $2 per watt.

The irregularity of solar irradiation on earth (day versus night, *Night* cloudy versus clear days, varying heights of the sun from zenith) has *and day* negative economic implications for photovoltaic generation, because of the large investment costs required for the photocells. The installed peak power capacity should, in fact, be four to eight times higher than the generated mean power.

These disadvantages could be avoided, if the cells were placed on
Sunny large satellite panels placed in geostationary orbits, at about 36 000-
satellites km height above earth, where there is the constant, full radiation of
1.35 kW/m^2. How to get the power safely and cheaply down to earth
is clearly the greatest problem to be overcome.

Photochemical processes

By far the most important photochemical process is the natural
photosynthesis which takes place in plants. This is the basis of the
biological cycle and thus of life on our planet. Natural photosynthesis
A flower occurs in the chlorophyll-containing chloroplast in plant cells. By
in your absorbing light energy, it takes up carbon dioxide (CO_2) from the
tank atmosphere and water from its surroundings and, after a complicated
process, turns these ultimately into biomass made of carbohydrates,
thereby releasing oxygen.

From an energy point of view one can distinguish two steps. The
first is concerned with the production of the biomass, through the
intensive cultivation of plants or efficient collection of wastes; the
second studies the transformation of this feedstock into fuels of
practical interest.[4.29] With regard to the latter, there are two main
processes, the biochemical one applied to wet biomass and the
thermochemical one (as gasification) preferentially fed by dry
biomass (Fig. 4.9). The biomass can be converted to practical fuels
such as methane, hydrogen, or alcohol; for example, ethyl alcohol
can be produced in amounts of about 300 liters per ton of biomass
made of mais (see Sec. 5.2 for applications).

Natural growing forests have a considerable potential for pro-
Trees in ducing the biomass. Down through the ages, wood has been, and in
the forest some areas still is today, the basic energy source. The annual net
primary production of the world's forests can be estimated as having
the energy equivalent of about 1.2 Q units, which amounts to about
four times man's present energy consumption, or to about 0.02% of
the solar energy reaching the earth's surface (Fig. 4.1).

Attention is paid today to intensive cultivation for biomass production. The dry biomass yield can amount to 30 to 60 tons per
Hyacinths hectare and year for eucalyptus trees, 40 t/ha·y for water hyacinth,
20 to 100 t/ha·y for algae, and 40 to 120 t/ha·y for sugar cane. It
should be noted, however, that the energy used to cultivate, fertilize,

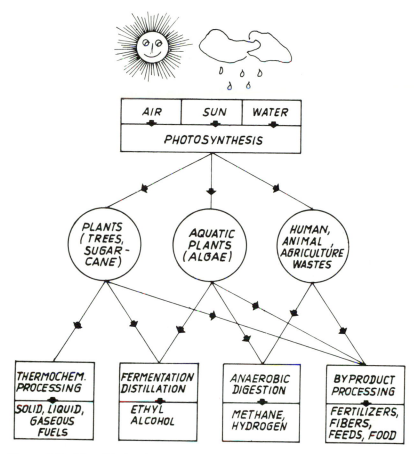

Figure 4.9 Simplified scheme of biofuel production from the biomass exploitation.

and harvest the plants in such cultivations can amount to about one-third of their energy content.

Studies are also in progress on more direct photochemical reactions. Of these, most attention at present is attracted by the production of hydrogen and oxygen through the solar light photolysis of water. Here again, many approaches are considered for this potentially very important process, but solutions are still far from being practical.

Heliothermal processes

The direct conversion of solar radiation into thermal energy is the most straightforward and simple application of solar energy. The central problem—not one of very advanced technology but of commercial costs—is the transformation of the low-temperature heat at the receiver (an indirect consequence of the low solar power density) into heat at high temperatures delivered at times appropriate for the end applications. This generally implies the concentration of energy, either by mirrors or (in the thermal cycle) by compressors, and its storage to cope with irregular insolation.[4.12]

Micro-processors All these steps have to be monitored continuously and operated according to the various needs of the users. This means that in most applications of solar radiation the problems lie not only with the single components but also with the overall system and its dynamic response to changing conditions.

As a rough classification, one can distinguish between low- and high-temperature heliothermal systems, not forgetting the indirect methods with their peculiar technical problems. Whereas heliothermal energy is usually exploited for simple heating, particular mention should also be made of its cooling applications. In fact, with regard to the latter, there is a near ideal coincidence between *The cooling* cooling requirements and maximum insolation. Cooling by solar *sun* energy can be achieved through the generation of electricity which then drives the air-conditioning units, or by the direct absorption (or other physical) methods that are used already in certain refrigerators.

Receivers that work through radiation-concentrating mirrors or lenses take advantage only of the direct radiation, and they must be constantly aligned with the sun. On the other hand, flat collectors or solar cells also take advantage of a part of the diffuse solar radiation.

It is interesting to note that a large variety of thermal solar systems have been in use over the last 100 years.[4.23] The largest application was a distillation plant designed by Carlos Wilson, built in 1872 in Las Salinas, Chile, and in operation for half a century, where the solar energy collected within a half a hectare still produced over 23 tons of fresh water daily for use in a nitrate mine. Flat plate collectors to drive an engine in a thermal cycle were used first by Charles Tellier in 1885 in France (20 m^2), later by Frank Shuman in 1908 in Pennsylvania (900 m^2), and together with

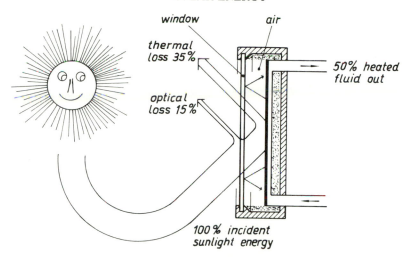

Figure 4.10   Low-temperature solar heat collector.

Note the window which is transparent to solar light, but opaque to most of the radiation of heat reemitted by the absorption surface (greenhouse effect). The collector thus acts as a trap for the solar radiation energy.

C.V. Boys in 1912 in Egypt; the type used by Boys actually consisted *Inventors* of mirror troughs in a 1 200-m^2 collector field that generated 50 kW *in Egypt* used to pump irrigation water from the Nile. These early applications of solar energy were often brilliant but too numerous to be remembered here. They were generally in use for short periods only, because of their uneconomical operation; the first solar pump was tested in 1875 by August Mouchot.

Low-temperature system. Direct application of solar energy for decentralized water and space heating is already widely in use. Even this simple application, where heated water is extracted mostly from flat collectors (Fig. 4.10), is at present energy cost levels (1982) only marginally, if at all, competitive. To guarantee regular delivery of heat for room and water heating beyond the normal storage time of a few hours, the solar system must be complemented by another energy source, such as an electrical heater. The efficiency of the associated thermal cycle is often upgraded by inserting a heat pump *Energy* which absorbs up to one-third of the energy in the cycle in the form *on a* of electricity. The successful application of decentralized solar *hot tin* energy for space and water heating thus often depends on the *roof*

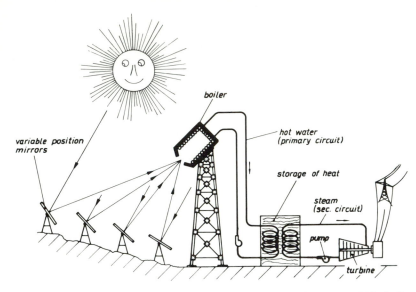

Figure 4.11   Schematic drawing of a high-capacity (over 5 MW) helio-thermal power station.

availability of sufficient and cheap electricity which happens to be one of the main goals of most electronuclear programs. Therefore, here is where nuclear and solar programs meet.

High temperatures. Some applications require heat at higher temperatures than those produced by the simple flat collector; for example, when the thermal cycle has to transform heat into mechanical energy and ultimately into electricity. The tower power plant (Fig. 4.11), where solar radiation is collected and concentrated by *Sunny* movable mirrors, is considered to be the most competitive system *towers* for the production of electricity at the multimegawatt level. The solar radiation is focused by a typical factor of 500 into the boiler, and the collected heat finally drives a turbine-generator set (Fig. 4.12). Of the various demonstration plants of this type in operation, the largest is the one near Barstow, California, with a total mirror surface of about 70 000 m² and a peak electrical power output of 10 million watts. Another is the Eurelios solar tower plant operated under the auspices of the Commission of the European Community near Adrano, Sicily; the plant has a total of 6 200 m² of mirrors and a peak electrical power output of 1 million watts; the working fluid

Figure 4.12 Aerial view of the thermal solar demonstration plant at Barstow, California, with a peak electrical power output of 10 million watts. (Courtesy: U.S. Department of Energy.)

is water and overheated vapor. Lower-capacity installations, in which this equipment would prove too costly, could fall back on concentrators based on fixed (or partly movable) spherical or parabolic mirrors offering concentrations in each single unit of up to a factor of 300.

Indirect thermal methods. The extraction of heat from seawater, lakes, rivers, and from the atmosphere by means of heat pumps (see Sec. 4.3) is an indirect way of harnessing solar energy. This method has already been applied successfully for many decades at the level of small or intermediate units. Nearly three-quarters of the earth's surface is covered by the sea which thus soaks up most of the solar radiation on earth like a giant sponge. It has always been tempting *Giant* to exploit this heat by massive ocean thermal energy conversion *sponge* plants (OTEC).[4.18] The first to consider the possibility of operating a heat engine with the warm ocean surface as a source of heat, and

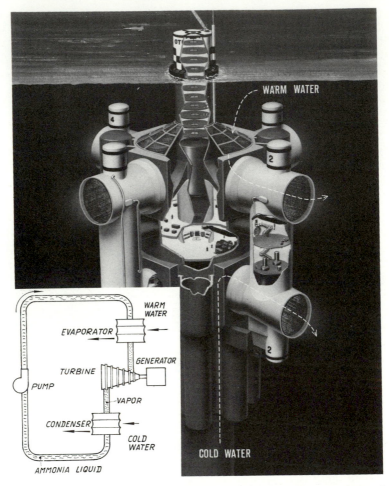

Figure 4.13 Artist's view of an ocean thermal energy conversion plant (OTEC). *Insert:* The temperature difference of about 20°C between the warm surface water and the cold water drawn from 600 m depth drives an ammonia-operated thermal cycle (Courtesy: Lockheed Corp., Sunnyvale, California.)

the colder, deeper ocean water as a sink was d'Arsonval in 1881; and in 1930 Georges Claude operated a prototype ocean power plant off the coast of Cuba, which generated a 22-kW gross power output. It is, however, only since the energy cost rose sharply that OTEC has been taken up again, with new projects that should be checked in demonstration plants from 1986 on. *Cuban connection*

Typically, such a thermal conversion plant is built on a platform anchored off the coast, where it operates by using the warm surface water to vaporize a working fluid, like ammonia, which drives a turbine-generator set (Fig. 4.13). The generated electricity could be used to produce hydrogen, ammonia, or methanol, or where possible transported to shore via a submarine cable. Meanwhile, cold water taken from a depth of around 800 m by a pipe perhaps 30 m in diameter for a 200-MWe plant is used to condense the vapor ready for the cycle to start again. The major drawbacks, once again, derive from the relatively low energy density actually usable from this source; the small temperature difference between the warm and the cold heat reservoir (about 20°C in tropical seas) results in a poor efficiency of the thermal cycle of less than 2%. Consequently, the installations are huge and require substantial use of energy-consuming materials. It was estimated that over a nominal life of 40 years of an OTEC plant, for every 10 J of useful energy produced, about 8 J had to be expended to build and operate the plant. *Floating energy*

4.5   ADVANCED NUCLEAR FISSION

While the debate continues as to the merits of the present generation of electronuclear plants, and of nuclear power altogether, development work is progressing on advanced nuclear reactor types along two basic lines of improvement, i.e., increasing the working temperature of the reactor and achieving self-sufficient breeding of the nonfissile natural uranium and thorium into fissile elements.

Both improvements, although being tested on various pilot or demonstration plants, still require substantial technological demonstration work on the reactor structures and on the related fuel cycles before they can become commercially competitive. Higher temperatures are achieved in various advanced gas-cooled reactors, with the bonus of increasing the efficiency of the thermal cycle; ultimately they might also allow direct application to chemical processes, such *Hot reactors* as the production of hydrogen, or to direct conversion into electrical energy through a process called magnetohydrodynamic conversion.

The breeder reactor is often characterized by simply saying that it produces more fissionable fuel than it consumes. In reality, by using excess neutrons during its normal operation, it transforms the non-*Breed* fissile abundant isotopes of uranium 238 or thorium 232 into fissile *and* plutonium 239 or uranium 233 that keep the reactor going, and it *multiply* also produces fissile fuel for additional reactors (Fig. 4.14).

The breeding process is assessed quantitatively by the breeding factor, defined as the average number of fissile nuclei produced by each fission event. In order to achieve an effective breeding with a positive fuel balance (i.e., to obtain a breeding factor larger than 1), a fission event must produce an average of more than 2.2 neutrons, because (see Fig. 3.1 and also the figure in Ref. 3.1) at least 1 neutron is needed to support the chain reaction, at least 1 neutron is needed to transform a fertile nucleus (uranium 238 or thorium 232)

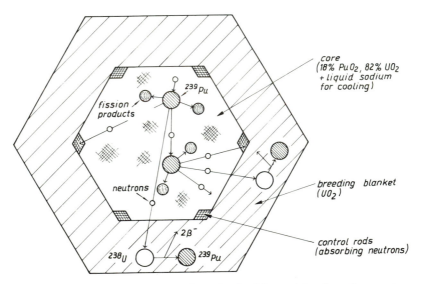

Figure 4.14 Schematic cross section of a liquid-metal fast breeder reactor. The chain reaction takes place in the active core of the reactor, mainly with plutonium 239. In the outer breeding blanket a nonfissile nucleus (uranium 238) is transmuted into a fissile one (plutonium 239). The plutonium thus produced is recovered chemically in reprocessing plants and supplied as an active fuel for subsequent recharging of the same or other reactors.

into a fissile nucleus (plutonium 239 or uranium 233); and at least 0.2 neutron on average is lost (e.g. captured in nuclei without giving rise either to fission or to plutonium transmutation).

The development of breeder reactors has so far been concentrated mainly on the uranium-plutonium cycle,[3.1] with liquid sodium as the coolant and without a moderator, where breeding factors of 1.1 to 1.2 can be obtained. This is the so-called liquid metal fast breeder reactor—LMFBR; see Table 3.1. A well-balanced system of such reactors and their associated fuel cycle (Fig. 3.5) could ideally in the long term consume 50 to 70% of the uranium initially fed into the cycle, thus improving uranium resources' exploitation by about a factor of 70, or more, with respect to exploitation in today's light water reactors.[4.9] In 1951 a reactor of this type (the American EBR 1) was the first in history to produce viable quantities of electricity. At that time it was conceived as an alternative option to the pressurized light water reactor for nuclear submarines. *First in history*

Improved uranium usage also means that ores with smaller uranium content can economically be exploited; the improved fuel cycle may thus finally lead to the exploitation of the enormous uranium resources which are contained in the seawater,[4.1] representing in practice an inexhaustible energy source (Fig. 4.2). The fission breeder reactor then will represent, by its own right, a renewable energy source.

Experimental fast breeder reactors have been successfully operated for several years, for example, the French Phénix. Out of this experience, the first demonstration reactor at the 1.2-GWe power level—Superphénix—has been built: it is scheduled to come on-load at Creys-Malville in the Rhône valley in 1985. The typical core of a 1-GWe fast reactor of the present generation contains roughly 4 tons of plutonium and 22 tons of natural uranium, with an additional 26 tons of uranium in the blanket (all in the form of oxides).[3.6] With a mean burnup in the core of 60 GW·day/t, the turnaround of the fuel is on a 2-year basis, with about 0.7 tons of fissile plutonium being burned per year. *Rising from the ashes*

Despite its undoubted advantages and successes, the large-scale introduction of the fast breeder reactor is not foreseen until after the turn of the century. Why? There are political, commercial, and technical reasons which can explain the relatively slow progress. One of the more important reasons is that the breeder reactor implies operation of the complete reprocessing cycle, where plu- *Advantages and doubts* tonium is extracted from the breeding material and used as new fuel. But, as we have mentioned in Sec. 3.2, considerable difficulties still lie in the way of reprocessing, even on a relatively small scale.

A second reason is that in order to obtain the fast neutrons, no moderator is used, and the reactor core is rather compact. The density of thermal power reaches values of 500 kW/liter, five times more than in the present generation of light water reactors. This puts great demands on nuclear fuel and cooling technologies.

Third, the nuclear industry is not ready to support the financial burden to develop and demonstrate the technologies necessary to bring the breeder reactor and its associated fuel cycle to the same degree of operational reliability as the one now obtained in present light water reactors. In fact, the nuclear industry is at present struggling for survival, and thus wants to continue to build the present generation of reactors for a few more decades in order to recover investments before facing the economic risks of the new breeder option.

Finally, the ease of breeding weapons-grade plutonium in the mantle and the large-scale amounts of plutonium circulating in the breeder's fuel cycle are considered by some a further contribution to the proliferation problems (Sec. 3.3).

The mentioned difficulties could certainly be tackled and solved within a short time if there was a real urgency in large-scale introduction of the breeder reactor (and the reprocessing of spent fuel, for that matter). Such an evolution could be favored, for example, by a substantial increase in price, or by a scarcity of uranium supplies. But this situation is not foreseen over the next one or two decades, and consequently the need for the breeder is at present not urgent at all.

An alternative to the fast breeder is provided by the uranium-thorium breeding cycle. Here the capture of relatively slow (quasi-thermal) neutrons by thorium 232 transforms this natural element into the fissile uranium 233 isotope.[4.9] This alternative cycle is *Thorium's* operated by reactors with moderators that are not very different *alternative* from those of present light water reactors, an important advantage over the fast breeder. But the major drawback of the thorium cycle in the long term is that breeding factors of only 0.9 are reached in practice, which indicates that self-sufficient breeding is not attained as a rule. In fact, these reactors are named converters rather than breeders.

Ultimately, the capability of breeding uranium and thorium into fissile isotopes is a question of having free neutrons available, easily and cheaply. A free neutron that breeds a fissonable isotope represents an energy bonus of 200 million electronvolts. Free neutrons are very precious, as 1 kg of them corresponds on this basis to an energy source of nearly 2×10^{16} J, or the equivalent of half a million tons of oil. Free neutrons were available in the Big Bang in the very first instances of existence of the universe (Sec. 2.1), when they were in reversible equilibrium with the intense radiation field. But neutrons could only survive later in the bound nuclear state (in deuterium first and in gradually synthesized heavier nuclei later), since a free neutron decays into a proton and an electron within a mean time of 11.7 min. *Free neutrons*

A unique source of free neutrons is the fission process and its multiplication through the chain reaction, as we have already seen. An important neutron source is also the nuclear fusion reactor (described in the next chapter), in which the nuclear reaction of a deuteron with a triton, or another deuteron, produces energetic free neutrons. *Proton connection*

A quite different neutron source which could become of practical importance is represented by an accelerator of charged particles. In fact, a single proton accelerated to high energies and then hitting a heavy metal target (e.g., uranium) can dislodge dozens of neutrons out of the heavy nuclei. Energetic neutrons, obtained by this "spallation" process or by nuclear fusion, can be further multiplied, since various elements (e.g., beryllium, lead, uranium) emit two or more neutrons when struck by an energetic one. As a result, one 1 000 million electronvolt proton hitting a uranium target can produce through spallation and multiplication typically 100 neutrons; and if only 70% of them induce fission events, the gross outcome is 14 000 million electronvolts of fission energy. In this process, therefore, the energy liberated amounts to 14 times the energy input by the proton. This represents an interesting energy amplification process, even if the various energy conversions (with their limited efficiencies) required to accelerate the proton, in the first place, reduce the energy gain in the overall balance.

CHAPTER 5

Energy: The Future

Nuclear energy—through fusion and fission breeders—, solar energy and, to a limited extent, solid fossil fuels are the conceivable solutions for providing the inexhaustible bulk of the very long term energy needs. Getting the energy from these and some other minor new sources and distributing it in socially and ecologically acceptable ways will become one of the crucial and challenging problems facing the future of our world. Speculating about future options and evolution is always a risky undertaking, but let us try. What are actually some of the technical and strategic options that will contribute to conceivably determining the future energy scenarios?

5.1 STRATEGIC OPTIONS AND CHOICES

Less than 50 years ago a walk on the moon was estimated to be 200 years away. Today it is history. What brought man to the moon was a clear decision, taken with determination, and built on a potentially mature technological base.

Walk on the moon There are many analogies and also some differences when considering the energy problem in comparison with the space mission. The technological base is available to exploit present and most future alternative energy sources. The decisions to proceed toward these goals are, however, not taken with a determination comparable to the one that sent a man onto the moon. Clearly, the energy problem is quite complex and strongly interrelated with economic and political facts, which make it more difficult—but not impossible—for clearcut decisions to be taken. The winning energy strategy at present seems to be "think globally and act locally," rather than to take big decisions.

Minerals from heaven and earth The energy strategies of the future have to be seen in the more general context of raw material resources,[5.26] of which energy is—so to speak—the most important one. In Chap. 1 the physical existence

118

of man was linked to the availability of energy, food, and materials, with the latter two strongly dependent on the first (energy and food, in fact, is the subject of Sec. 1.2).

The continued large consumption of certain minerals—particularly those containing metals, as for example iron and uranium[4.1]—will gradually shift the prospection to ore deposits with smaller and smaller mineral content; extraction will thus gradually become more energy (and cost) intense.[5.25] Although these interrelated aspects involving the dynamic concepts of reserves, resources, and costs (see also in Sec. 4.1 and 4.3) deserve careful analysis, it can be said that shortage of minerals (and metals) will not be a substantial problem of the future, given the availability of energy as indicated in Chap. 4. In addition, recycling of metals, exploitation of the resources in the oceans and eventually in the outer space, and the shift toward nonmetallic materials (although again energy related) will favor the availability of metals.

In recent years, many critical and often pessimistic analyses of the energy future have been published. In 1972, the Club of Rome's "The Limit of Growth" warned of possible social and ecological disasters "within the next 100 years." More recently, the "Global 2000 Report to the President," published as an official U.S. government document,[5.15,5.12] warned that "life for most of the world's people...could...be more difficult and more precarious in the year 2000 than it is today." These and other such reports *Bad* most often are formally correct and cautious with their forecasting, *messages* but their messages generally reach the public as bad news only. At this level, the energy problem looks nearly always catastrophic.

The negative tilt in communicating with people seems to be a general effect for which some reasons should exist. For example,

- Bad news sells books and newspapers, whereas good news is not half so interesting.
- Bad news points to problems that should be solved, thus requiring programs to be established and people to be hired.
- Some people seem to be attracted by the cumulative nature of exponential growth or multiplication by large numbers with negative (or positive) signs only, rather than by the selective, but tedious balance of many negative and positive numbers or facts.

In drawing scenarios on the future energy needs,[4.11,5.15] one has to consider both the evolution of the per capita consumption and of the population on earth.[5.12,5.18] To reach the first 1 billion around 1830, homo sapiens required about 40 000 years, and for the second billion only 100 years. The third was reached 30 years later (1960) and for the fourth, 15 years were required. Today (1984), there are 4.76 billion people on our planet, with a present increase of 1.7%, or 80 million, per year. Although there is reason to believe that in future the population explosion will be contained, nevertheless it is *Crowded* estimated that there will be 6 billion people by the year 2000 and *world* maybe 8 billion 20 to 30 years later, with a final stabilization at around 10 to 11 billion at the end of the century.

On the other hand, the per capita consumption in countries that evolve from the preindustrial to the industrial phase was illustrated and discussed in Chap. 1 (Fig. 1.1). Most likely, the evolution curve depicted in this figure will apply also in future. A major question that remains is related to the evolution of the per capita consumption in countries that enter the postindustrial phase. In fact, it is generally *Less* accepted that the per capita consumption can and will diminish, *is* the only question being how much.
enough

A look at what has happened in the past could help in understanding the future. World energy consumption in the past 100 years has on average increased at the rate of an overall 2% annually. This figure, however, was 3% for the United States, making it the biggest energy consumer in the world. (It should be noted, on the other hand, that energy consumption per inhabitant has always been high in the United States; in 1850 it was just under half of today's figure.) The energy consumption of the industrial countries in the 1960s rose by approximately 7% annually, but from 1973 onward the average increase was only a few percent, and at present, in the industrialized western nations, it is slightly decreasing. This welcome slowing down of consumption was caused by the massive energy cost increase, the economic recession, and a greater sensitivity toward conservation. But for the poor and developing countries, the increase in the energy use must continue, if they ever intend to catch up with the industrialized world.

So what will the world's probable average consumption of energy be in the year 2000? Table 5.1 illustrates schematically why the world will then probably be consuming about 1½ times the energy it consumed in 1984. At the 12th World Energy Conference in Delhi (1983) it was estimated that the (optimistic) consumption of energy in the year 2020 could amount to about 18 billion tons equivalent

Table 5.1   Consumption forecast

Year	Population (billions)	Energy requirement (million equivalent barrels/day)	Observations
1984	4.7	140	Of which 85 are oil and gas, 40 coal, 15 other[1.17]
2000	6	180	With mean per capita consumption as in 1984
		210	Increased by one-sixth to take into account the needs of the third world

petroleum (corresponding to 370 million bbl/day; see Table 5.1). At *Out to the* the Conference in Istanbul (1977), by comparison, the same figure *future* was about twice as high, showing how in these years the estimates have become more conservative.

The only major options foreseeable today for providing the bulk of the long-term energy needs (in 50 to 100 years' time and onward) are nuclear energy through fusion and fission breeders, solar energy and, to a limited extent, coal. The intrinsic drawbacks of solar energy (low-energy density, intermittent availability, high costs; see Sec. 4.4) mean that even in the long term its share will remain *Fusion* limited, say within 25% of the overall energy production, a substan- *and* tial fraction of this being for the so-called low-temperature thermal *fission* cycle exploitation. Coal should, at best, have a similar share, as higher values are unlikely on account of the major ecological re-percussions which this energy source involves (Chap. 1).

If the previous reasoning applies, and when the time comes to substitute most oil and natural gas, nuclear energy should likely cover over 40% of the world's total energy requirement. The ques-tion as to which specific source of nuclear energy will predominate in the medium and long term remains open. Nuclear fusion, des- *Open* cribed at the end of this chapter, is potentially a very interesting *question* solution in the long term, but its technological feasibility and its economic competitiveness remain to be proved. Consequently, an intellectually interesting and practically important question is: When will nuclear fission energy be necessary? Is it at all necessary?

The facts and considerations presented on nuclear energy in Chap. 3 and Sec. 4.5 have shown that nuclear energy already provides a considerable share of electricity, has been safe, and (not counting the huge investment in research and development in this sector by the military in the last 45 years) is economically competitive. In addition, the solution of major remaining problems, e.g., waste disposal, high level safety, must be started anyway in the next 20 years in support of the existing civilian and military nuclear plants.

Coal or not coal

Nuclear fission energy is a large-scale technology, whose major application is, and will remain, the generation of electricity, for which application it should meet the increased electricity demand, gradually substitute oil, and limit the use of coal. About 25% of the world oil consumption (700 million tons in 1982) is burned in electric power stations. As a 1-GW (electric) oil-fired power plant burns in the mean about 1.4 million tons of oil per year, this substitution alone would require about 500 large nuclear power plants. In any case, the confirmation of nuclear energy strongly depends also on the success and further expansion of electricity, an evolution that seems likely and which will be discussed in the following section.

It is likely that within the next 30 to 50 years:

- Coal will not cover more than 25% of the total energy demand (ecological implications, uneven distribution among nations).
- Solar energy will also be limited in its penetration (high cost, material intensive, storage problems).
- Oil and gas together will not increase much above the present overall output (limited reserves, strategic thinking).
- Electricity will predominate as an energy vector, absorbing about 40% of the primary energy for its production (practical, efficient, allows small-scale technological solutions at consumer level).

Desirable fission

The logic answers to these questions on a global basis would seem to be that nuclear fission is highly desirable, if not clearly necessary. If so, then the building of nuclear power plants should be taken up again on the short to medium term, as soon as nuclear energy has been more favorably accepted by the majority of people. Nevertheless, the absolute necessity of fission energy does not follow from this reasoning alone, as one could change the assumptions to suit an energy scenario without nuclear energy. But would this exclusion be realistic and convenient?

The general considerations made so far on a global scale should really be referred to the conditions and necessities of single continental areas. The United States, for example, with their vast coal resources[4.6] could well do without nuclear energy, while western Europe's energy production would be predominantly nuclear.

The long-term success of nuclear fission energy, say from the year 2030 onward, depends on the large-scale introduction of the breeder reactor, as discussed in Sec. 4.5. At that point, nuclear fusion energy could gradually emerge as a real alternative energy option to the fast breeder.

It goes without saying that considerations on the energy strategies of the future depend on a much more complex and rich matrix of options and possibilities. Central to any energy policy in the industrialized world is conservation. The extended use of small-scale technologies for efficient energy application and conservation at or near the final consumer level is potentially another important option, as it could markedly influence the overall policy. *Complex matrix*

The introduction of any new technology takes time to mature and to gradually assess its economic potential, in terms of both financial and energy revenues. One speaks in terms of penetration times, which for the most successful examples of new energy sources lie typically between 30 and 50 years. The penetration difficulties of a bright idea into the harsh market realities of today, and the effort and time it takes to overcome them, are another important consideration to be made in evaluating future energy policies. It took just 10 years for the electric generation and distribution system to spread throughout the industrialized world after Thomas Edison patented the incandescent lamp. Today, the relative penetration times can be longer, but then technologies are more complex and generally the market size, and thus its inertia, is much larger. The electric utility industry has become so large that its sheer scale inhibits rapid penetration. *Penetration to success*

Generally speaking, a new technological idea must mature through three or four stages before reaching widespread commercial use.

- The first stage concerns the proof of the idea's scientific feasibility, that is, demonstrating that the concept works in a laboratory environment.
- The second stage is to demonstrate engineering feasibility, that is, translating the concept into practical and progressively larger

devices: from a pilot plant (just large enough to incorporate key subsystems) to a medium-sized demonstration plant (with most of the subsystems scaled up to near commercial size), and finally to a fully commercial-size demonstration plant.

- Commercial competitiveness is the third stage; it really includes two important and critical substages: commercialization and economicity. In the former, private or public investors begin to place orders and insert the technology into their systems. However, before a new technology gains a significant space in the market, it must prove to be really economically competitive with alternative solutions.

In this evolution process, each stage is substantially more expensive and risky than the previous one, and the process of decision making—particularly if it is a purely civilian, nonmilitary enterprise—

Business and politics

becomes ever more demanding. In this context one would expect that for the not too complex technologies, the economic competition brings each single stage, and the work done therein on subsystems or the overall system, into a proper ordering with society's priorities, and with its political realities. However, both business people and politicians generally expect return benefits from their investments or involvement on time scales of only a few years. This leaves great problems for the public sector (and of some big companies) to guarantee an efficient and successful management of the large technology programs which often take 20 to 30 years, or more, to develop and deploy (see Sec. 5.3 for nuclear fusion energy as one extreme example).

Chianti and champagne

Take the example of nuclear power and the jet aircraft. Both concepts were demonstrated in 1942: The first controlled fission chain reaction in the Fermi pile, and the first U.S. military jet flight by the experimental Bell aircraft. In these two cases, thanks to an unprecedented military-sponsored development effort, the engineering demonstration was achieved in the United States in the following 10 years. Within less than another 10 years the third stage was reached: the first commercial jet passenger service (the British Comet, 1957; the U.S. Boeing 707, 1958) and the first commercial nuclear plants online (Dresden 1 and Yankee Rowe in the United States, 1960). From this stage the penetration speed was very different for the two cases: Commercial jet traffic had reached 25% in 1963 and a near 100% in 1968, whereas nuclear penetration reached 10% of electric power production only in 1975. In contrast to these two examples, the relatively simple, but important transistor took just 5 years from invention to its entry into the marketplace.

To conclude this intriguing and complex section with a certain *Imaginative* order, we sketch one of the possible energy scenarios and group it *scenarios* along three successive phases.

The first 25 years (up to the year 2010). The world's energy consumption will rise steadily, reaching globally near to twice the values of 1984. Practical convenience, and shortsightedness about the strategic and ecological value of the most precious natural gas and oil, will further increase their consumption. Nevertheless, toward the end of the period, at least one-fifth of the added energy capability should be met by nuclear power stations of the light water type, based on well-established technology. This implies that a satisfactory solution to the problem of reprocessing and/or storage of the nuclear wastes will have been defined on an international basis by then. Electricity production capacity should be more than twice the present one, so as to require about one-third the overall primary energy consumption.

The following 40 years (from 2010 to 2050). Nuclear high-temperature reactors and breeders (based on the plutonium cycle or the thorium cycle) will gradually become competitive and will play an integral part in the production of nuclear energy, which could be supplying over one-third of the world's energy requirement by the end of the period. Solar energy (hydro energy excluded) will perhaps account for somewhat less than 10% of total consumption after the year 2030. During the same period, some thermonuclear fusion power stations might also be operational, even if not yet economically competitive. Extensive use of the electric car and of the heat pump will further increase the relative share of electricity to about 40% of the total primary energy consumption. Energy conservation through district heating, heat pumping, and the recovery of discharge heat will alone or in interconnected arrangements substantially improve the efficient use of energy for heating, thereby saving energy equivalent to about 20% of the total consumption.

The future beyond the year 2050. The burning of oil and natural gas as a source of (heat) energy will substantially diminish, and later disappear, thus radically changing the structure of the energy economy. (The residues in oil and gas wells still accessible to the much advanced drilling techniques will be used as feedstocks for the manufacture of synthetic materials, and as a practical and readily available energy store.) Hydrogen and other synthetic fuels produced in nuclear power stations will be in common use. The structure of the energy market will be determined only in part by the momentary cost of the involved energy sources, as ecological and strategic considerations at the international level will finally become predominant. The world consumption of energy in these circumstances might be kept down to four to six times the present volume. Nuclear fission and fusion energy might provide over one-half the energy requirement, while solar energy might supply 20% or more.

Such an assessment of the medium- and long-term prospects in the energy context is obviously very much a matter of speculation (albeit based on a series of recent international studies), and it therefore invites a critical and imaginative analysis of the organization and the needs of the future human society.

5.2 BOTTLED ENERGY

For the practical use of energy at the consumer level, the energy of primary sources is generally transformed into various "fuels." In addition to transportability, safety, and easy applicability, the practical value of these fuels also depends on their storage capability; in this respect, the stored mechanical energy in a spinning wheel[5.8], or the electricity stored by electrolysis in a battery, can also be considered a "fuel." In Table 5.2 different forms of fuels are grouped

Fuel fantasies into four categories; chemical, nuclear, mechanical, and electrical. Nowadays, most energy at the level of the individual consumer is used in liquid form (as a derivate from crude oil), or as natural gas, or in the form of electrical energy.

The phasing out over the next 50 to 100 years of the compact fuels derived directly from oil and natural gas requires a systematic study to identify the best substitutes for storing and distributing energy for practical purposes such as transport and domestic heating. This general problem entails an adaptation of energy-consumption practice, which generally presupposes an intensive material (i.e., energy)

Energy vectors investment to start with; for example, small-scale solar energy exploitation in remote and poor countries, or the efficient waste heat recuperation and transport for district heating in the industrialized world. More in general, the problem requires the identification of the energy vectors between the major sources of the future

Table 5.2   Possible classification of some "fuels" used for the practical storage and release of energy

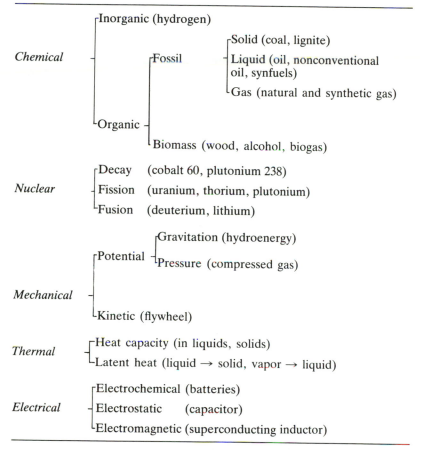

(nuclear fission and fusion energy, solar energy, and coal) and a multitude of small- and medium-sized users.

One of these vectors is electricity which has for a century proved its unique practicality and flexibility. The production of electricity is peculiar in that it relies on very large scale technology in central *Large and* power plants, such as nuclear plants; but then, at the user's end, *small* electricity also favors the efficient application of small-scale energy technology, such as the practical exploitation of solar thermal energy.

Electricity is here to stay and will further expand its share in overall energy consumption, probably absorbing nearly 40% of the primary energy in the long term. As we have previously discussed substantial increase of the thermoelectrical cycle's efficiency can be obtained with available technologies at all its stages—production, transport, and usage. At the first stage, for example, with cogeneration, at the last with heat pumps that efficiently use solar energy *No* and low-temperature ambiental heat. The losses in the transport and *storable* distribution systems of electricity, which amount at present up to *fuel* 20%, can also be reduced. In long-distance transport this could be achieved by switching toward the million volt transmission lines,[5.10] from the present 300 000 to 400 000 volt, or by using superconductors.[5.9] Further network integration and a tight computerized control can improve distribution.

The lack of direct and convenient storage possibilities is one of the main disadvantages of electricity. In fact, large-scale storage must be done via conversion to mechanical or potential energy, and reconversion to electricity when this is required; for example, in large pumped-storage hydroelectric plants where electricity is used to pump water uphill, from which it is then reconverted by hydro-electric generators when needed; or in the compression of gas in large underground caverns, from where it is released and expanded in gas turbines to drive electrical generators. On the small scale, there are various electricity storage possibilities, but they are still relatively expensive and impractical.

Energy and environmental aspects would make it particularly important to have in future electricity accumulators (batteries) for driving private transportation means competitively in comparison *Dream* with other fuels (Fig. 5.1). In practice, the electrical car does not *car* exist today because the available batteries are too heavy; they are also too expensive and their lifetime is too short. It is generally expected that this situation will not change in practical terms for at least the next 20 years.

The lead-acid battery is the most widely adopted at present in power applications; it was introduced 100 years ago by the French chemist Gaston Planté. Good units can attain a stored energy of 30 Wh/kg of weight or of 80 Wh/liter of volume, and live through 500 full 80% discharge-charge cycles (gasoline's energy equivalence is 12 900 Wh/kg). In practical terms, the energy stored in a fully loaded battery can barely bring an equivalent volume of water to ebullition. A minimal goal of competitiveness for application in

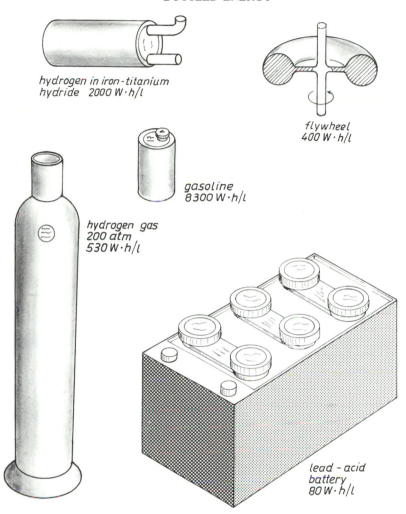

hydrogen in iron-titanium
hydride 2000 W·h/l

flywheel
400 W·h/l

gasoline
8300 W·h/l

hydrogen gas
200 atm
530 W·h/l

lead - acid
battery
80 W·h/l

Figure 5.1 Forms of stored energies of interest for private and public transportation.

The indicated energy densities are expressed in watt-hours per liter of volume (1 Wh corresponds to 3600 J); notice for comparison that the energy required to heat 1 liter (or 1 kg) of water from 0 to 100°C is 72.5 Wh, whereas the heat for melting at 0°C is 93 Wh and for evaporation at 100°C is 630 Wh. The volume of the symbolized storage containers are for an equivalent amount of stored energy, with the volume increased by a factor of 3 for the gasoline and hydrogen fuels; this because the energy efficiency in converting their thermal energy in mechanical energy for propulsing the vehicle is about 22%, i.e., about three times worse than for the battery or flywheel.

transportation requires for a battery at least 200 Wh of stored electrical energy per kilogram, and a lifetime of more than 1 000 full charging-discharging cycles, at reasonable costs. This performance would make the electrical car interesting for use at least in urban areas, as it should allow a range between recharges of 150 km and a lifetime of the batteries expressed in 50 000 km traveled distance (the battery should, however, also be able to drive the car at a maximum cruising speed of 80 to 100 km/h). Of the various batteries under study today (nickel-zinc, zinc-chloride, lithium, and lead-acid types, etc.),[5.11] the sodium-sulfur type comes nearest to these specifications; but it must be operated at a temperature of about 400°C and requires further technological developments.

Chemical plug In addition to the electrical batteries just mentioned, which are charged by plugging them into an electrical source, there are cells which generate electricity by the influx and reaction of chemical elements.[5.21] Such fuel cells are not new; more than 140 years ago the scientist William Grove described the concept. Fuel cells operating on the combination of hydrogen with oxygen into water and producing electricity are used in space exploration. The fuel cell has numerous attractions as a source of heat, electrical power, or both; since there are no moving parts, it operates noiselessly and fumelessly and can attain energy efficiencies of 80%. But, unfortunately, the cells are still expensive since they need catalysts of precious metal to run efficiently.

Nuclear hydrogen The proven practicality of liquid and gaseous fuels would seem to point to the replacement of natural oil derivatives or of gas with synthetic fuels produced from the primary energy sources of the future;[5.20] for example, nuclear energy used to produce hydrogen directly or via electricity. Other possibilities are synthetic fuels produced by coal hydration, or ethanol produced directly from the fermentation of biomasses (Sec. 4.4).[4.28] Gasohol (gasoline diluted with 5−10% ethanol and with no polluting lead compounds in it) is already available in many countries; it can be burned in any gasoline-powered automobile.

Sweet gasoline The corn or mais required to replace 1% of gasoline with alcohol in the United States could provide the 2 500 food kcal per capita required to sustain about 15 million people. The explicit cultivation of biomass for energy production points once again toward the strong interrelation between food and energy. In fact, this practice could require land otherwise used for food cultivation,

and thus may in certain poor countries seriously limit food production and increase food prices. This, clearly, is not what a global
energy policy wants to achieve.

Hydrogen would seem to be a particularly attractive synthetic fuel
by virtue of its flexibility, its comparative ease of storage, its ease of
transport through pipelines, the previously mentioned possibility of
directly producing electricity in fuel cells, and its non-polluting properties[5.7]. Regarding the last quality, it can be noted by looking at
Fig. 2.9 that its combustion with the oxygen contained in the air just
produces water (although some traces of polluting nitrogen oxides
will inevitably be formed when using air to "burn" the hydrogen).
These properties have been demonstrated thanks to the extensive
use of hydrogen in the space programs. The attractive potential of
this fuel, however, may have exaggerated the prospects of rapid and *Cold*
widespread application. More critical considerations show that the *fuel*
energy content of hydrogen is low when compared with conventional
fuels (Fig. 5.1). A car running on hydrogen must carry nearly
120 liters of hydrogen in the most condensed, liquid form, whereas
it would carry only 40 liters of gasoline which, moreover, is easier to
transport and store.

Hydrogen can be stored as a gas, as a liquid, or integrated into a
solid. Hydrogen in the gaseous form has the lowest density, even if
compressed at up to 200 atm. Hydrogen in the liquid phase is
obtained only at the extremely low temperature of minus 253°C by a
liquifaction process which requires a consumption (i.e., loss) of up
to 60% of the hydrogen's energy content of 2 400 Wh/liter (for
liquified natural gas, the loss is only about 15% as a consequence of
the more convenient temperature at which it is obtained).

Storing hydrogen as a solid-bound compound is, by comparison,
simple, safe and energy-efficient[5.5]. The storage is obtained by having
hydrogen gas flowing over an appropriate metal compound where it
may be bound by chemical reaction to form a metal hydride. This
compound is then slightly heated to liberate the hydrogen for usage.
The amount of hydrogen stored per unit volume in this way can be
higher than in the liquid phase: for example, iron-titanium hydride
stores 96 g of hydrogen per liter, whereas the liquid has only 70.8 g. *Powerful*
On the other hand, because of the weight of the metal compound, *sponge*
the hydrogen per unit weight of the store compared with the liquid
phase is 60 times less for the iron-titanium hydride and 15 times less
for the more favorable magnesium hydride. In fact, a distinct advan-

tage of pure hydrogen as a fuel is its low weight. Liquid hydrogen has an energy content of 33 000 Wh/kg, 2½ times as much as gasoline which has 12 900 Wh. This represents an important advantage for aerospace applications even if oxygen (in a quantity eight times heavier than hydrogen) must be carried along for the combustion process (usually just the oxygen from the air is taken).

Most hydrogen is produced today by chemically processing natural gas (a process termed reforming or cracking); similarly, hydrogen can also be obtained from hydrocarbons. Other methods do not use fossil fuels as raw material and thus relate conveniently to the *Wet* nuclear or solar primary energy sources.[5.16] One is by direct elec- *energy* trolysis of water, the reverse of the reaction shown in Fig. 2.9: Through the input of electric energy, hydrogen and oxygen are produced out of water.[5.19] In practical terms only about 60% of this energy is recovered as thermal energy when hydrogen and oxygen recombine at the user level (a maximum of 85% could be reached in the future). Since electricity itself is produced with 30 to 40% efficiency from a primary thermal energy source, such as from nuclear energy or from fossil fuels, the overall cycle efficiency remains less than 25%.

5.3   FUSION ENERGY FOREVER

The controlled nuclear fusion process involving the two isotopes of hydrogen constitutes at present one of the most important research projects on the energy of the future.[5.27] Over and above its major practical objective—i.e., energy—it also represents an intellectual *Intellectual* challenge in the shape of a formidable task of pure and applied *power* research, and technological development. The scientific and technological difficulties, which must be solved before the commercial application of nuclear fusion might become possible, are still paramount, but the final goal is more than tempting—it is the realization of a new, inexhaustible energy source. For these challenging aspects and their strategic importance, the problem of nuclear fusion is dealt with in some detail in the concluding section of this book.

Controlled nuclear fusion

The nuclear fusion reactions between some of the lightest nuclei[2.1] represent together with gravitation the fundamental energy liberating process in the universe. Actually, as most of the condensed mass is still in the form of protons (either free or bound with an electron into a hydrogen atom), fusion fuel is just about everywhere in the universe. Fortunately, however, the reaction among two protons is extremely difficult to obtain, since otherwise the evolution of the universe would have been a totally different, violent event. *Violent universe*

The extremely slow and difficult synthesis of protons into deuterons and, after some intermediate steps (see Ref. 2.1) into helium, remains the smooth energy source of the stars. Also, the energy produced in the sun—and radiated into the space at a continuous power level of 3.8×10^{26} W—comes from the fusion of the proton fuel contained in its central core and heated at a temperature of about 15 million Kelvin. The sun is a daily reminder that thermonuclear fusion works and delivers energy (Sec. 4.4). *Heavenly source*

Under the ambiental conditions of the earth, proton nuclear burning is plainly impossible; otherwise, the oceans (because of the proton content in the water molecule), if they existed at all, would have represented a tremendous block of fusion explosive. But fortunately (as has been discussed in Sec. 2.1), in the course of the cosmological evolution, our planet has been provided with a nuclear fusion fuel that is up to 1 billion times a billion (10^{18}) times easier to burn: It is principally deuteron (the nucleus of the deuterium atom) which is also present in seawater in the proportion of one deuterium for every 6700 hydrogen atoms. On this basis the fusion energy content in 1 liter of seawater is equivalent to the energy in 300 liters of gasoline (see Table 4.1), and thus, by all practical means, fusion represents an inexhaustible energy source for man (Fig. 4.2). It is because of the tiny, but significant content of deuterium in seawater that nuclear fusion is sometimes dubbed the "method for burning the seas." *Burning seas*

A large number of exothermic fusion reactions between the lightest elements available on earth are known and studied in the laboratory.[5.1] For practical reasons, only the reaction among the nuclei of the two isotopes of hydrogen, deuteron and triton, and the nucleus of helium 3 are worth considering. In view of the practically inexhaustible, nonradioactive deuterium reserves on earth, the most *Inexhaustible reactions*

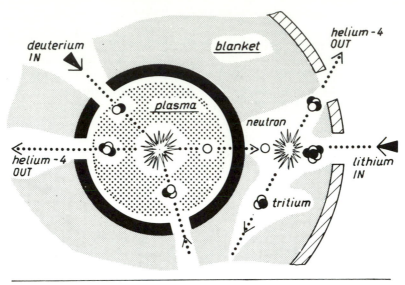

$$\text{fusion} \;:\; \text{deuteron} + \text{triton} \rightarrow \text{helium} \cdot 4 + \text{neutron} + 17.6\,\text{MeV}$$

$$\text{breeding}: \; \text{neutron} + \text{lithium} \cdot 6 \rightarrow \text{helium} \cdot 4 + \text{triton} + 4.8\,\text{MeV}$$

$$\text{balance}: \; \text{deuteron} + \text{lithium} \cdot 6 \rightarrow 2 \times \text{helium} \cdot 4 \quad\quad + 22.4\,\text{MeV}$$

Figure 5.2   Deuteron-triton fusion reaction in the hot thermonuclear mass (plasma) and triton breeding reaction in the outside lithium blanket (the reaction energies are given in MeV).

straightforward reaction is the fusion of two deuterons (see reaction 1 in Ref. 5.1). However, experiments have shown that the fusion reaction of a deuteron with a triton (Fig. 5.2) can occur roughly 100 times more easily in the types of fusion reactors considered up to now. As even with the deuteron-triton reaction the practical exploitation of fusion turns out to be extremely complex, the more difficult straight deuteron reaction will have to wait quite a long time before becoming of any practical importance.

In all these fusion reactions, the liberated energy is carried away in the form of kinetic energy by the particles that result from the reaction process; only subsequently can the fusion energy be transformed into radiation or heat by collisions of the particles with other nuclei or electrons. Of the large deuteron-triton fusion energy (17.6

million electronvolts), a full 80% is carried by the emitted neutron, with the rest going into the kinetic energy of the helium nucleus. With these facts in mind, one expects (Fig. 5.2) that a deuterium-tritium fusion reactor contains a mass of reacting nuclear fuel surrounded by a blanket of inert material where the easily escaping neutrons deposit their energy by collisions, while the electrically charged helium nucleus with its kinetic energy is more likely retained in the reacting mass (particularly if there is a magnetic field, as will be described further on). *Energetic carriers*

Since tritium is radioactive with a half-life of 12.26 years—decaying by emission of an electron with a mean energy of 5.7 thousand electronvolts into helium 3 (this corresponding to a radioactivity of 10 000 curies per gram)—it must be created artificially. The most convenient way is to breed it by neutron capture from either one of the two lithium isotopes;[5.2] the breeding logically would take place in the lithium-containing blanket (Fig. 5.2). The original elements thus fueling the deuteron-triton reaction are deuterium and lithium; both are nonradioactive. The terrestrial energy resources represented by this fuel cycle, even if smaller than for deuterium only, are again inexhaustible by any practical measure (Fig. 4.2). *Tritium breeding*

In relation to these fundamental and interesting aspects concerning fusion, it is not surprising to learn (see Sec. 2.3) that it was the study of the light nuclei's reaction processes that led to discovering nuclear energy in the 1930s, even though it was nuclear fission which then produced the first nuclear energy with a practical meaning. During the Manhattan project (the development project of a nuclear fission bomb during World War II), the problem of igniting a thermonuclear fusion process (through a fission detonator) was discussed on several occasions and studied particularly by E. Teller at Los Alamos. *Lively discussions*

At the same time, work at Berkeley under the leadership of E. O. Lawrence on the mass spectrographic method of separating uranium isotopes implied a thorough experimental and theoretical study of the plasma formed in electrical arcs. The confinement conditions of hot plasmas by magnetic fields (the state of the matter in which fusion takes place, as will be explained later) was thereby developed more clearly than had ever been done before. It is thus understandable that after the war, lively discussions and simple experiments occurred in various laboratories on basic problems related to the containment and heating of plasmas of thermonuclear interest.

A new phase in the search for ways to tame the thermonuclear flame began with the large research programs set up in the United States, in the Soviet Union, and in the United Kingdom around 1952. These programs were spurred on by the successful effort that led to the first (explosive) ignition of a thermonuclear fuel on earth *Atoms* at Eniwetok Atoll on November 1, 1952. In opening the first Atoms *for* for Peace conference at Geneva in 1955 the late Homi Bhabha, the *peace* Indian president of the Conference, astonished the public opinion, which was not aware of these ongoing fusion programs, by predicting that a method would be found for liberating fusion energy in a controlled manner within the following two decades. After the disclosure in 1956 of these research activities (which had already identified most of today's schemes and approaches to controlled thermonuclear fusion), many new fusion programs started all over the world, notably in most of the European countries.

The second Atoms for Peace Conference in 1958, with its unprecedented release of information on fusion research and its historical display of fusion devices in the U.S. pavillon, contributed to bolster-*Fusion* ing up hope in the new energy source. This optimism, however, *pavillon* soon dissipated somewhat as the physics of thermonuclear plasma revealed its intrinsic complexity. Controlled thermonuclear fusion research has evolved since then into an exciting scientific activity that in its presently most studied configuration—the tokamak— offers the realistic chance of demonstrating, within a decade, the ignition and containment of a fusionable plasma.

The controlled release of fusion energy on earth remains a most *Challenging* challenging undertaking, and actually is the subject of one of the *research* largest research and development programs of nonmilitary, world-wide interest. In 1984 various programs (including magnetic and inertial containment fusion) occupied approximately 7000 professionals (scientists and university engineers) and required financing of over $1800 million, of which the European community's program contributed with about $400 million, and the United States $600 million.

The thermonuclear process

The question arises of how to obtain net energy in a practical reactor operated on the basis of the fusion reactions already mentioned. In the case of nuclear fission the solution happened to be simple, the trick being the chain reaction in fissile uranium 235, demonstrated for the first time in the Fermi pile in 1942 (Sec. 2.3). For nuclear fusion the problem is quite different and more difficult.[5.3] But the stars and the sun show vividly that one possible solution does exist: it is the thermonuclear process, which consists of

Shining promise

- Heating a mass of fusionable elements to ultra high temperatures (above 100 million Kelvin for a deuterium-tritium mixture) so that the thermal random motion becomes sufficiently violent to have the colliding nuclei undergo fusion with a finite, though small probability
- Containing this hot mass during a time long enough for fusion reactions to take place and to produce a useful output of energy

So defined, the nuclear fusion process looks formally similar to chemical combustion, as in fact it is. A minimum temperature is also required to ignite any chemical reaction process, which then proceeds and expands ("burns") to produce net thermal energy, if the reacting mass is held together long enough. But because nuclear energy density is 1 million times higher than chemical density, the time scales, the propagation velocity, and the ignition temperature are completely different.

A sort of fire

At the high temperatures required for nuclear fusion, the heated atoms that compose the fusionable mass break up into their constituents, the electrons and nuclei. These particles form a plasma (Fig. 5.3) and move around in what is called their thermal random motion; actually, the higher their temperature, the higher their velocity. In their movement, the particles are continuously exchanging and sharing their kinetic energy through collisions among themselves, and thus tend to have the same energies, or velocities, a situation defined as thermodynamic equilibrium. However, as this is a random process, there are always some particles that by chance happen to have temporarily a kinetic energy much above the mean (thermal) kinetic energy of most of the particles in the hot mass.

Hot collisions

It is just up to such a very small fraction of particles that move

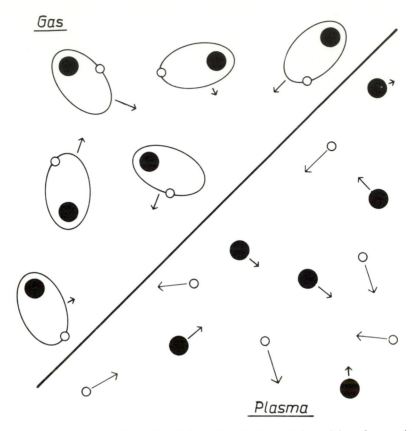

Figure 5.3   Atoms (or molecules) are the fundamental particles of a gas; in a plasma, on the other hand, electrons and nuclei (or ionized atoms) move separately. A plasma is obtained by heating the gas to sufficiently high temperatures, for example, through an electrical discharge.

around much more violently than the other particles to be able to fuse in a collision (obviously, the fusion reaction happens only in the collision of two nuclei, as the electrons at these energies do not react). For example, in a deuterium-tritium mass at 100 million Kelvin, the fusion reactions are mostly carried out by the nuclei that have an energy three times as high as the mean thermal energy (of 9 000 electronvolts at this temperature), representing only a very small fraction of all the nuclei in this mass.

The thermonuclear approach to fusion (which is at the basis of practically all present-day fusion experiments) is thus very inefficient; *Inefficient* a lot of energy must be spent to heat all the particles just to have *Approach* very few of them at a high enough energy to induce some fusion reactions. In the stellar environment and time scale such consider- ations are clearly of little value, particularly if thermonuclear fusion, once ignited, proceeds in an autocatalitic way.

What about more efficient approaches to ignite and sustain a nuclear fusion process? The answer is that there are no other obvious or simple solutions; for example, just accelerating some nuclei to high enough energy and shooting them into a fusionable mass will not provide useful fusion energy. Probably one way to improve fusion's chances of success could be to invent ingenuous means for slightly distorting the thermodynamic equilibrium long *Ingenuous* enough to have more high-energy nuclei, and thus more fusion *distortion* output, on the basis of the same heating requirement. These re- marks are made in support of the opinion that progress in controlled fusion depends not only on improvement in technology and en- gineering, but eventually also on fresh, new physics ideas.

The methods used in present-day experiments to heat the fusion- able mass to temperatures of tens of million degrees will be described further on. With regard to containment, the necessary condition can actually be expressed quantitatively by requiring that the confine- ment parameter, i.e., the product

Nuclear density × energy confinement time

(in other words, the number of fusionable nuclei per unit volume n *Breaking* times the time τ it would take to loose half of the thermal energy *even* from the contained plasma) be larger than a critical value. For a given composition of the fusionable mass, this value depends on the temperature and on the level of the energy balance required.

This $n \times \tau$ confinement condition is defined in Fig. 5.4 for an ideal deuterium-tritium mixture. The condition for "energy break- even" is obtained when the overall energy balance in a fusionable mass is positive, i.e., if the total fusion energy produced during the contained phase (17.6 million electronvolts per deuteron-triton reaction; see Fig. 5.2) is greater than the energy required initially to heat the mass and to make up for the inevitable losses through radiation, particle migration, or heat conduction.

If, instead, one considers only the energy deposited locally within the plasma by the reaction products (only the 3.52 million electron-

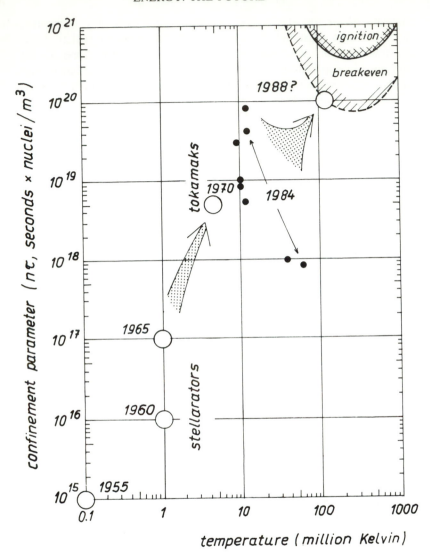

Figure 5.4   The figure depictes some of the confinement parameters and corresponding temperatures obtained in past and present experiments; all experiments strive toward the indicated region—the fusion heaven—where breakeven or ignition conditions are achieved in a deuterium-tritium plasma.

volts deposited by the helium nucleus, neglecting the additional energy carried away by the neutron), the "ignition condition" is obtained. As a consequence of this local energy deposition, the thermonuclear plasma will increase its temperature, thereby augmenting its reaction rate, and thus finally ignite, similarly to what can happen to a chemical reaction that propagates at ignition. It is common to refer to the ignition condition as the condition of scientific feasibility of controlled thermonuclear fusion, although often the less severe breakeven condition is indicated as the next significant milestone to be achieved by fusion research programs.

With regard to plasma containment, there are in principle various solutions, as shown in Fig. 5.5. Gravitational confinement (a) functions in the sun, but it cannot be reproduced on our planet because gravitational forces are too weak. The very high temperature of a thermonuclear plasma also rules out nearly completely the more immediate solution of a solid wall vessel (b), since this would immediately be vaporized and destroyed. The other two solutions, inertial and magnetic confinement, are being experimentally ex- *Opposite* ploited. They are pursuing opposite solutions to obtain the necessary, *solutions* minimal value of the $n \times \tau$ confinement parameter: relatively low densities (n) and thus lengthy times (τ) in tens of seconds for magnetic confinement; very large densities (thousand times the density of the solid material) and thus extremely short times for inertial confinement.

Inertial confinement fusion actually consists of microexplosions of deuterium-tritium pellets, with dimensions of a few millimeters (Fig. 5.6). Inertial confinement is, in fact, based on the inertia of the fuel mass, which resists somewhat the natural expansion when it is very rapidly heated to thermonuclear fusion temperatures, thus formally providing the necessary "confinement times" of a few *Contained* billionths of a second. Although it is conceptually a simple and *explosions* interesting method, the microexplosion approach, because of its large technical difficulties, is behind the quasi-steady, magnetically confined solution in demonstrating the feasibility of nuclear fusion (this will be described in the next section). In this there is a certain parallelism with the development of the thermomechanical combustion engines. The steady steam engine—for two centuries the only method for converting heat into mechanical energy—preceded the explosive, internal combustion motor which in many areas of applications turned out to be the most successful solution.

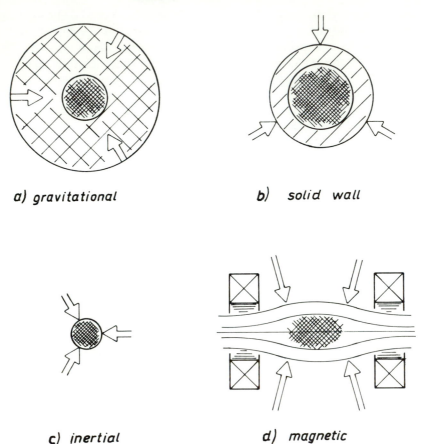

a) gravitational   *b) solid wall*

c) inertial   *d) magnetic*

Figure 5.5   Thermonuclear plasma confinement systems: (a) Gravitational, (b) solid wall, (c) inertial, (d) magnetic.

Some important considerations must be made in connection with the physical state of matter heated to very high temperatures in which the thermonuclear process takes place. As the mass is heated to temperatures above about 100 000 K, the electrons in the hydrogen *Hot plasma* atoms free themselves from the nuclear attraction and the gas becomes a plasma which is composed of electrons (possessing a negative electrical charge) and the hydrogen nuclei (with positive charge). In addition, one speaks of ions which, when one or more electrons are detached from the atoms, also remain positively

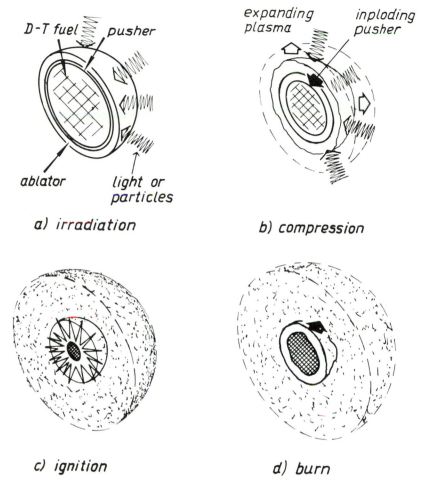

a) irradiation

b) compression

c) ignition

d) burn

Figure 5.6 Operating principle concerning a microexplosion of a spherical pellet, upon which inertial confinement fusion is based.

The predicted sequences encompass the following stages: (a) Laser light or particle beams rapidly heat the ablator surface that surrounds the spherical pellet; the latter consists of a spherical shell (the "pusher") which encloses the deuterium-tritium fuel. (b) The rocketlike, violent blow-off of the rapidly heated ablator material causes the implosion of the shell by mechanical reaction. (c) Through the pusher shell implosion, the fuel core can reach 100 to 10 000 times liquid density and ignite at a temperature of 100 million Kelvin. (d) Thermonuclear burn spreads rapidly through the compressed fuel, while mounting pressure reexpands the pusher and fuel, thereby terminating the reactions before fuel is burned completely; nevertheless, energy output could amount to 100 to 1 000 times the energy input through laser light or particle beams.

charged. In a plasma, the ions (or nuclei, in which case the atoms are completely stripped from all the electrons) move separately according to their own kinetic motion (Fig. 5.3).

The interaction between atoms in the gas is the result of direct knock-on collisions; in a plasma, the interaction occurs by means of the electromagnetic fields of the electrically charged particles in motion, the effects taking place even at a relatively great distance, with many other particles (only the stronger and more direct im-

Collective pacts achieve contact between nuclei and therefore fusion). This
effects fundamental characteristic gives rise to collective interaction effects in the plasma, which in turn support an extremely broad spectrum of oscillations. These effects become even more complicated when there is a magnetic field that influences the motion of the charged particles (Fig. 5.7). Consequently, the physical properties of a plasma composed of charged particles are much more complex than those of gases.

The presence of free electrons (as in a metal) entails a high electrical conductivity of the plasma, which increases as the temperature rises; for example, a plasma at 100 million Kelvin conducts 30 times better than copper at ambient temperature. Plasma is defined as the fourth state of matter after the solid, liquid, and

The fourth gaseous states. Nearly all matter in the universe takes the form of
state plasma. On earth plasma has to be generated artificially. It occurs, for example, in flames, flashes of lightning, neon tubes.

Studying the physical characteristics and properties of plasmas is a fundamental part of fusion research. Plasma physics integrates all disciplines of classical physics; it is an ideal medium to study many large amplitude oscillations and nonlinear phenomena. The understanding of plasmas—which is of interest to astrophysics and to many areas of technology and industrial applications—is benefiting immensely from the research effort made within today's fusion programs.

Magnetically confined reactor

A hot plasma can be isolated from solid walls and contained by a magnetic field. This is because its constituents—the electrically

Spiraling charged ions and electrons—remain "bounded" to the field lines as
paths they follow spiraling paths along these lines (Fig. 5.7). This property

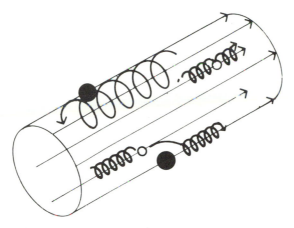

Figure 5.7   Containment of charged particles by a magnetic field.
Particles are spiraling along the magnetic field lines but can jump from
one line to another if disturbed. In the figure, an electron (open circle)
moves nearby an ion and is deflected by attraction through its electric field
onto a new field line.

gives rise to the magnetic confined fusion concepts which represent
the main line of research in today's fusion programs.[5.27]

In relation to the most powerful mean magnetic fields, which can
reasonably be generated in conventional coils (expressed in units of
6 to 10 tesla, i.e., about 100 000 times stronger than the mean
terrestrial magnetic field), the density of the deuterium and tritium
fuel is limited to that of a gas 100 000 times less dense than normal
atmospheric pressure (i.e., 3×10^{20} particles per cubic meter),   *Burning*
which technically corresponds to a good vacuum. If it were denser,   *vacuum*
the pressure exerted by the corresponding plasma, when heated to
thermonuclear temperatures, would overcome the macroscopic
confinement capability of realistic magnetic fields: in a way, the
magnetic field containment would explode like an overheated tire.
With these densities the ignition condition for deuterium-tritium
fuel shown in Fig. 5.4 then requires confinement times longer than
tens of seconds.

Unfortunately, magnetic confinement is never perfect (Fig. 5.8).
Loss of energy from within the plasma and outward to the containing walls can occur through radiation, migration of its particles, or a
heat conduction process analogous to that in a metal. As implied by

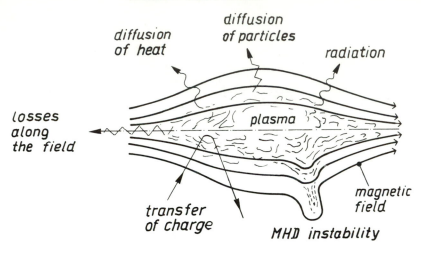

Figure 5.8   Some of the most significant losses of energy and particles in magnetic confinement.

The conduction of heat and the diffusion of particles is dominated by anomalous processes due to various microinstabilities. In addition, during the process of charge transfer, a neutral atom enters the plasma and loses an electron to a hot ion which can then move unhampered through the magnetic confinement. The loss through irradiation includes bremsstrahlung, and recombination and cyclotron radiation.

Fig. 5.7, charged particles move much more easily along the magnetic field lines than across them. In order to at least avoid the large losses along field lines, the simple open magnetic field configuration is bent to meet the ends into a closed, ring-shaped ("toroidal") configuration. To satisfy basic equilibrium and stability conditions of the enclosed plasma, it is actually required that the field lines encircle the toroidal plasma in a screwlike ("helical") pattern: By going around the torus many times, the magnetic field lines define a closed surface, thereby containing the toroidal plasma (Fig. 5.9).

Hot doughnut

With the two closed (toroidal) configurations of the so-called stellarator and tokamak type, an enormous progress in magnetic confinement fusion has been achieved during the past 30 years: the confinement parameter has improved 100 000 times, and the temperature nearly 1 000 times (but not on the same experiments, as shown in Fig. 5.4).

A remarkable improvement in these parameters was obtained

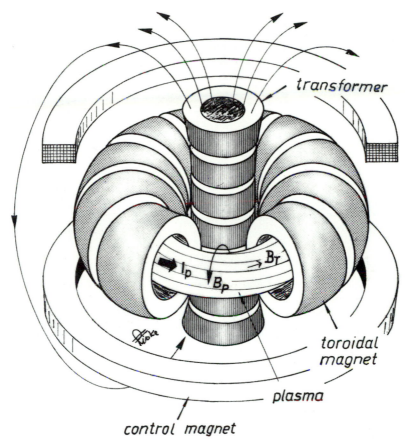

Figure 5.9 Diagram of a toroidal confinement system of the Tokamak type.

The transformer induces the electric current I_P which heats the plasma and generates the poloidal magnetic field B_P. The main magnet produces the toroidal field B_T. Also shown are the coils which produce a control field used to center the toroidal plasma in the discharge chamber. The poloidal and toroidal fields combine into the helical field pattern required for equilibrium and stability reasons.

Figure 5.10 Internal view of the toroidal vacuum chamber of the Joint European Tokamak within which a thermonuclear plasma is generated and heated to temperatures aiming at the 100 million Kelvin level.

When generated, the plasma made of electrons and hydrogen nuclei is kept at the center of the chamber, away from its wall, by magnetic fields generated through coils located on the outside; it is heated to temperatures of thermonuclear interest by a combined action of a large electrical current flowing around the chamber, and electromagnetic waves and particles injected through the portholes in the walls, for a final total power of about 25 MW. (Courtesy: JET Joint Undertaking, Culham.)

Three successful coils

toward the end of the 1960s with the successful application of the tokamak configuration. (Tokamak is an acronym from the Russian "*to*roidalnj *ka*mera *mak*ina"—machine with toroidal chamber). It consists of three sets of magnetic coils, which enable a toroidal plasma to be generated, contained, and heated within a toroidal vacuum vessel (Figs. 5.9 and 5.10). Today more than 70% of the worldwide fusion effort is devoted, directly or indirectly, to the tokamak configuration which is producing the best-known plasmas in fusion research. For this reason the new generation of large experiments, which will come into full life in the second half of the 1980s, are nearly exclusively of the tokamak type, such as JET[5.13] (the Joint European Torus) at Culham near Oxford and the Tokamak Fusion Test Reactor at Princeton University in the United States. One expects that these experiments will toward the end of the eighties produce deuterium-tritium plasmas at up to, or near to, breakeven conditions, thus possibly demonstrating the scientific feasibility of nuclear fusion.

How, actually, are these plasmas heated to near thermonuclear temperatures?[5.28] After having been formed from a gas through an *The* electrical discharge, the plasma in a tokamak is heated at first by the *chamber's* strong electric current induced along the toroidal plasma by the *heating* transformer (this heating is based on the same principle that turns an incandescent light bulb on: the plasma's electrical resistance will dissipate as heat the energy of the current that is forced to flow in it).

Above about 20 million Kelvin the resistance heating becomes inefficient because at that temperature the plasma's resistance starts to be negligible. Additional heating methods applied to boost the temperature toward the 100 million degrees level involve injecting into the plasma electromagnetic energy at a broad spectrum of radio frequencies and/or high energy beams of deuterium and tritium atoms. In JET, for example, the additional heating power will total 25 million watts.

Notwithstanding the success of the tokamak, it is generally considered necessary and wise to pursue research and development work on other configurations or ideas as well, so as to create or *Stars* preserve alternative options. One is the inertial confinement concept *and* mentioned earlier (Fig. 5.6); other options refer to open or to *mirrors* toroidally closed magnetic confinement configurations (mirror machines or stellarators). Even with respect to the tokamak, there is work along different lines of approach. An example is the very high magnetic field tokamak that promises to attain near ignition conditions in a more compact device than would be possible with a low field JET-type concept. But such a compact tokamak is not directly extrapolable to long-life reactor operation, because of the extremely large mechanical loads and high radiation power densities to which its components are exposed.

Although the specific physics problems of containing and heating thermonuclear plasmas are as yet not fully solved, the basic design

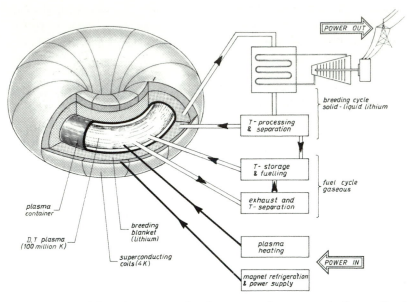

Figure 5.11   Major elements and subsystems of a magnetically confined fusion reactor based on the deuterium-tritium (lithium) fuel cycle (full lines indicate power transmission; open lines indicate particle or fuel flow).
A thermal power of 6 000 MW (i.e., about 2 300 MW electric) implies a daily consumption of 0.7 kg of deuterium, and 1 kg of tritium or an effective 8 kg of lithium; the total tritium inventory in the system would be 10 to 20 kg. For a reactor operating with a mean magnetic field of 5 tesla, the minor and major radii of the toroidal plasma are typically 2 and 6.4 m, surrounded by a blanket and coil structure at least 2 m thick.

and operating conditions of possible fusion reactors, as one can foresee them today, are relatively well established. Unfortunately, *Big* the magnetically confined, toroidal fusion reactor just "does not like *and* to operate at small sizes." Typical data qualifying such a reactor *complex* points toward a large and complex energy system that will require the development and testing of many subsystems and new technologies before it can be put into operation (Fig. 5.11)[5.29].

Magnet technology provides some of the basic components of any magnetic fusion device. To avoid the enormous resistive losses of *Dropping* large magnets with pulse lengths of hundredths of seconds, in pre-*resistance* sent conceptual designs it is generally accepted that at least the main (toroidal) magnet must be superconducting. Thus work in this area is devoted to the development and application of superconductivity, a particular phase of some metallic alloys (as niobium-titanium) where the electrical resistivity at very low temperatures drops practically to zero.

One of the main technical characteristics of the tokamak-type fusion reactor is that thermal energy recovery and breeding of tritium from lithium takes place in the blanket surrounding the toroidal discharge chamber.[5.2] While this feature is an advantage from the point of view of the dynamics and control of the reactor, it also entails certain drawbacks. In particular, the wall of the discharge chamber (the so-called first wall) is subjected to an intense bombardment of particles and radiation, since through it the energy *Active* is conveyed from the thermonuclear plasma to the blanket (80% *blanket* being carried through the wall by fast neutrons). For this wall and important parts of the blanket to survive 10 and more years of operation, new materials are required with outstanding characteristics against radiation damage, chemical corrosion, and many other unique properties.

The technology of the tritium fuel is mainly concerned with demonstrating the reliable and safe handling of the radioactive tritium stored and flowing in the reactor system. More in general, it will be *Handle* important that the development and demonstration of fusion-related *with* technology be accompanied by the continuous assessment of the *care* safety built into the reactor system and the environmental impact of fusion energy.

In conclusion, fusion energy is characterized by exclusive properties, some of which represent distinct advantages over the other *Exclusive* major energy sources. They can be grouped around three points: *outlook*

- *Fuel*. Inexhaustible supply of cheap fuel (deuterium and lithium) which can be found virtually anywhere; it is nonradioactive, and its collection (mainly from sea water) causes insignificant ecological problems.[4.2]
- *Burn*. Fusion reactions produce energy and no direct radioactive waste with all its problems (no afterheat, no waste disposal). However, in presently contemplated fusion reactor concepts there is radioactivity from two sources. The first is tritium, which is produced locally by breeding from lithium, then extracted, purified, and "burned" directly.[5.2] The second is activation of the reactor structures by neutrons not used up by the breeding. Future reactor concepts may strongly limit this radioactivity.

- *Output.* Fast neutrons, charged particles, and radiation carry the energy out from the reactor core, in shares dependent on the used fuel.[5.1] These carriers allow some specific, unconventional applications of fusion energy, other than thermal cycle exploitation for electricity production (as shown in Fig. 5.11), such as:[5.23] direct magnetohydrodynamic conversion of heat into electricity; synfuel production through photochemical processes; efficient reduction of ores; transmutation of fission wastes into more convenient radioactive products; breeding of fissile fuels.

Sophisticated engineers Notwithstanding all these exciting promises of fusion, there must be no illusions about the technical difficulty, the cost, or the time, likely to be required to bring even the deuterium-tritium (lithium) fuel cycle to a successful demonstration level. In fact, the relative complexity and large size of currently conceivable fusion reactors, the very stringent requirements on materials for their structures, and the severe operating conditions which the electromechanical components are subjected to during operation, will require an unprecedented contribution from the most sophisticated engineering and from advanced technology.

Cheaper than ships Even if such a deuterium-tritium reactor were ever built around today's conceptual designs, probably no one would want it because of its complexity and cost. On the other hand, there is no specific reason today to doubt that finally fusion could be made practical and successful. Now and again, major technological projects (those not hampered by scientific limits) have passed successfully through the nearly inevitable, evolutionary simplification that accompanies general technological progress and man's determination and ingenuity—and have finally reached a breakthrough. Who would have believed 70 years ago, with flying already an exciting reality, that planes would finally provide economic transport of passengers across the Atlantic—and be cheaper than ships?

Mass appeal In nuclear fusion (in the deuteron-triton reaction or in the full deuteron-deuteron cycle) four-thousandths of the mass entering the reaction are transformed into energy, four times as much as in nuclear fission. Complete annihilation (transformation) of matter into energy brings a further gain of a factor 250.

Annihilation and also nonthermonuclear fusion are currently induced in the laboratory through the catalytic action of artificially generated elementary particles. For example, when a 200 times heavier, negatively charged μ-meson particle substitutes the electron in the normal proton-electron hydrogen (see Fig. 2.10), the linear *Fictional* size of this muonic hydrogen atom shrinks by a factor 200. This *epilogue* effect can be used to confine a muonic molecular ion made of a deuteron and a triton with one bound μ-meson in such a small volume that there will be a sizeable chance of a spontaneous fusion process between the two nuclei even at below room temperatures. Recent experiments have shown that one μ-meson, during its short lifetime of about 2 microseconds, can induce up to 180 nuclear fusion processes in a deuterium-tritium mass; and more could be possible in the future[5.30]. If muonic fusion could be transformed into an autocatalitic energy process, where new μ-mesons would be created by a fraction of the liberated fusion energy (more than 100 million electronvolts of energy are required alone to create the meson's mass), then it would represent an obvious advantage over the relatively inefficient thermonuclear process with its required very high temperatures.

Complete annihilation is obtained in the laboratory when antimatter is brought in contact with normal matter, for example, an antiproton with a proton or an antielectron (positron) with an electron. But in our world antimatter does not exist freely, and (as it is known today) these antiparticles must be generated artificially in large particle accelerators through processes which are very energy-inefficient.

Ideas about practical energy generation with these two methods, or other processes involving artifical elementary particles, pertain at present to the domain of science fiction. But for how long?

References

Chapter 1

1.1 G. Leach, in *The Man-Food Equation*, A. Bourne (ed.) (Academic Press, New York, 1973): The energy costs of food production. J. S. Steinhart and C. E. Steinhart, Science *184*:307 (April 1974): Energy Use in the U.S. Food System. G. M. Ward, Th. M. Sutherland, and J. M. Sutherland, *Science 208*:570 (May 1980): Animals as an energy source in third world agriculture. See also Ref. 1.10.

1.2 World Development Report 1980 (World Bank, Washington, D.C., 1980).

1.3 J. Darmstadter, J. Dunkerley, and J. Alterman, *Annual Review of Energy 3* (1978): International variations in energy use: Findings from a comparative study. See also Ref. 4.14.

1.4 H. Rose and A. Pinkerton, *The Energy Crisis, Conservation and Solar* (Ann Arbor Science Publ., Ann Arbor, Mich.), p. 83.

1.5 E. J. Wasp, *Scientific American*, p. 42 (November 1983): Slurry pipelines.

1.6 J. B. Fenn, *Engines, Energy, and Entropy*—A Thermodynamics Primer (Freeman, San Francisco, 1982).

1.7 Neue Zürcher Zeitung: Vorzüge der Pipelines als Transportmittel (June 23, 1981); Alaska Gasleitungsprojekt um Meilen weiter (May 28, 1981).

1.8 W. R. Murphy and G. McKay, *Energy Management* (Butterworths, London, 1982). E. L. Harder, *Fundamentals of Energy Production* (Wiley, New York, 1982). K. O. Thielheim, *Primary Energy—Present Status and Future Perspectives* (Springer-Verlag, Berlin, 1982).

1.9 J. A. Fay, *Technology Review*, p. 51 (July 1983): Harnessing the tides.

1.10 R. C. Fluck and C. D. Baird, *Agricultural Energetics* (AVI Publ. Co., Westport, Conn., 1980).

1.11 Proceedings Nashville Conf. *Health Impact of Different Sources of Energy* (IAEA, Vienna, 1982). E. E. Pochin, *Physics Technology 11*:93 (1980): Biological risk involved in power production.

1.12 N. Smith, *Wood: An Ancient Fuel with a New Future* (Worldwatch Inst., Washington, D.C., 1980)

1.13 W. A. Nierenberg (chm.), *Changing Climate* (National Academy Press, Washington, D.C., 1983).

1.14 G. H. Kinchin, Risk assessment, in W. Marshall (ed.), *Nuclear Power Technology*, vol. 3 (Clarendon Press, Oxford, 1983).

1.15 W. O. Alexander, in *Evaluation of Energy Use*, Report No. 6 (The Watt Committee, London, 1979): Total energy content and costs of some significant materials.

1.16 Data on Gross Domestic Product (GDP), in OECD, National Accounts of OECD Countries, Vol. 1, Main Economic Indicators.

1.17 Data on World Energy Consumption, in OECD, Energy Balances in OECD Countries; ENI, Energia e idrocarburi; BP—Statistical Review of the World Oil Industry. (See also Ref. 5.17.)

1.18 R. S. Berry, T. V. Long, and H. Makino, in *Energy Analysis*, A. G. Thomas

154

(ed.) (Ref. 1.19): An international comparison of polymers and their alternatives. *Making the Most of Materials* (Science Res. Council, United Kingdom, 1977). See also Ref. 1.15.

1.19 A. G. Thomas (ed.), *Energy Analysis* (Westview Press, 1977)

Chapter 2

2.1 *Major synthesis reactions in the universe.* The formation of the universe, in particular of the stars, is characterized by a large number of synthesizing nuclear reactions that under particular circumstances gradually form heavier and heavier elements, up to iron and beyond. The most significant reactions are grouped into some basic sequences in the following table. The symbols used are: γ, gamma radiation; v, neutrino; e^+, positron, i.e., positive electron; p, proton; n, neutron; D, deuteron; T, triton; He, helium; C, carbon; Be, beryllium; N, nitrogen; O, oxygen; Ne, neon; Fe, iron; U, uranium.

n. Reactions	Remarks
Proton-neutron reaction in Big Bang	
1 $p + n \rightarrow D + \gamma$	The most likely elementary synthesis, as p and n are about equal in numbers
Helium synthesis in Big Bang	The predominant helium-producing reactions in Big Bang
2 $D + D \rightarrow {}^3He + n$	
3 $D + D \rightarrow T + p$	
4 ${}^3He + D \rightarrow {}^4He + p$	
5 $T + D \rightarrow {}^4He + n$	
6 $4p + 4n = 2{}^4He + \gamma$	Balance of helium production in Big Bang
Proton-proton reaction in stars	Only helium-producing reaction in stars up to temperatures of 15 million K. At 10 million K, reaction 7 takes more than 10 billion years mean reaction time.
7 $p + p \rightarrow D + e^+ + v$	
8 $D + p \rightarrow {}^3He + \gamma$	
9 ${}^3He + {}^3He \rightarrow {}^4He + 2p$	
10 $4p = {}^4He + 2e^+ + 2v + \gamma$	The energy balance in reaction 10 is $2 \times 1.44 + 2 \times 5.49 + 12.85 = 26.71$ MeV $= 4.27 \times 10^{-12}$ J from which the neutrinio's energy, 2×0.26 MeV, is sometimes subtracted.

Carbon cycle in stars	Predominant helium catalyzing reaction in stars for temperatures larger than 15 million K.

11    $^{12}C + p \rightarrow \, ^{13}N + \gamma$
$\overline{\qquad\qquad\qquad\qquad}$
                      etc.

First reaction of a successive proton capturing chain involving the isotopes ^{13}C, $^{13,14,15}N$, ^{15}O

12    $^{12}C + 4p = \, ^{12}C + \, ^{4}He + 2e^{+}$
      $+ \, 2\nu + \gamma$

Energy balance as for No. 10, but neutrino's share is 1.7 MeV.

Triple-α process in stars

Helium burning at temperatures larger than 150 million K. Most important step in the synthesis of heavy elements. Requires a dense core.

13    $^{4}He + \, ^{4}He \rightarrow \, ^{8}\overset{*}{Be} + \gamma$
14    $^{8}\overset{*}{Be} + \, ^{4}He \rightarrow \, ^{12}C + \gamma$

Three-body process, as $^{8}\overset{*}{Be}$ decays in 2×10^{-16} s if no further helium is captured.

15    $3^{4}He = \, ^{12}C + \gamma$

16    $^{12}C + \, ^{4}He \rightarrow \, ^{16}O + \gamma$

The inner core of massive stars contains carbon and oxygen in roughly equal amounts.

Iron synthesis in heavy stars

17    $^{12}C + \, ^{12}C \rightarrow \, ^{20}Ne + \, ^{4}He + \gamma$
$\overline{\qquad\qquad\qquad\qquad\qquad\qquad}$
                          etc.

18    $2^{12}C + 8^{4}He = \, ^{56}Fe +$
      $+ \, 2e^{+} + 2\nu + \gamma$

Carbon fusion reactions begin in a very dense core. The process ends at a temperature of about 5 billion K as the iron (and some chromium and nickel) does not deliver any energy if it reacted further.

Neutron capture in supernovae

19    $^{56}Fe + n \rightarrow \, ^{57}Fe + \gamma$
$\overline{\qquad\qquad\qquad\qquad}$
                      etc.

Core triggers supernova explosion by gravitational collapse. Heavier elements are formed predominantly by successive neutron capture in explosion and the following phases of stellar evolution.

2.2 H. Reeves and J.-P. Meyer, *Astrophysical Journal 226*:613 (1978): Cosmic ray nucleosynthesis and the infall rate of extragalactic matter in the solar neighborhood.

2.3 *Nature*, vol. 300, p. 312 (15 November 1982). Th. Gold and S. Soter, *Scientific American*, p. 130 (June 1980): The deep-earth-gas hypothesis.

2.4 E. Teller, *Energy from Heaven and Earth* (Freeman, San Francisco, 1979)

2.5 D. O. Hall and K. K. Rao, *Photosynthesis*, 2d ed. (Arnold, London, 1976).

2.6 G. Holton (ed.), *American Journal of Physics 49*:205−231 (1981): Hystory of the atom. S. Weart, *American Journal of Physics 45*:1049 (1977): Secrecy, simultaneous discovery, and the theory of nuclear reactors.

2.7 L. Spitzer, *Searching Between the Stars* (Yale University Press, New Haven, Conn., 1981).

2.8 Le Phénomène d'Oklo. Compte rendu d'un colloque tenu à Libreville, Juin 23–27, 1975 (IAEA, Vienna, 1975).

Chapter 3

3.1 *Fission process.* The basic fission and breeding reactions are collected in the following table, whereas the graphs in the figure express quantitatively some of the fundamental physical properties that characterize the various types of nuclear reactors in dependence of the neutron kinetic energy. The latter is expressed in electron volt units (eV, see Table 1.1), to each value of which there is obviously a corresponding velocity of the neutron, quoted separately in kilometers per second. For example, a neutron having a kinetic energy of 1 million eV has a velocity of 14 km/s.

Table to Ref. 3.1:   Some reactions using heavy nuclei. Meaning of the symbols: γ, emission of γ rays; β^-, emission of electrons; n, neutrons.

Nuclear fission of uranium 235, 233, and plutonium 239

- ^{235}U (or ^{233}U or ^{239}Pu) + $n \rightarrow$ on average 1.5 to $2.5n$ + 2 medium-weight nuclei + 200 MeV

Breeding of natural uranium 238 and thorium 232

- $^{238}U + n \xrightarrow{\gamma} {}^{239}U \xrightarrow{\beta^-} {}^{239}Np \xrightarrow{\beta^-} {}^{239}Pu$ (plutonium cycle)

- $^{232}Th + n \xrightarrow{\gamma} {}^{233}Th \xrightarrow{\beta^-} {}^{233}Pa \xrightarrow{\beta^-} {}^{233}U$ (thorium cycle)

In particular, the curves on part (a) of the figure give the number of neutrons emitted for each neutron absorbed by the various isotopes indicated, in dependence of the energy (or velocity) of the primary neutron producing the fission event. For example, a 200 000-eV neutron, having a velocity of 6.2 km/s, produces on average 1.9 neutrons when absorbed in uranium 235 and 2.4 neutrons when absorbed in plutonium 239.

The many neutrons in a nuclear reactor move with widely different velocities, i.e., they are described by an energy (or velocity) spectrum. Already at their birth during the fission event their velocities are different, as shown in the figure. This fission spectrum shows, for example, that the most frequent neutron energy is about 1 million eV and, also, that the number of neutrons emitted at around 1 000 eV is lower by factor of 20.

3.2 *Characteristics of reactor core.* For the PWR-fuel one considers a thermal energy production (burnup) of 33 GW·day/ton of fuel (corresponding to 800 GW·h/t) and a reactor load factor of 0.8; one-third of the fuel charge must

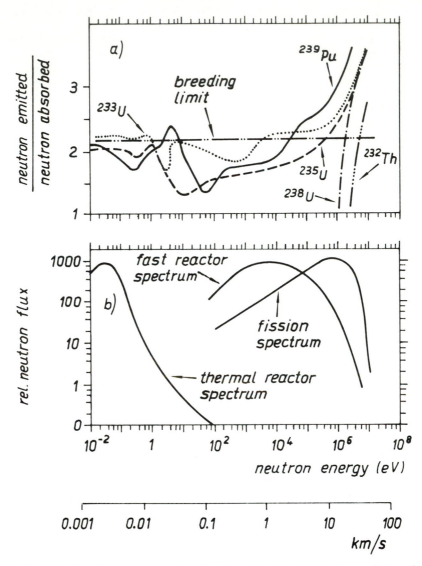

Figure to Ref. 3.1: (a) Variation of the number of neutrons emitted by fission of various isotopes per neutron absorbed, in dependence of the neutron energy. (b) Energy spectra of neutrons emitted by the fission process, and typical neutron spectra in a thermal and in a fast reactor.

be renewed yearly and then stays in the reactor for a 3-year cycle (Table 3.2 actually describes the products relative to this fuel portion at the end of the cycle). Other reactor types have different fuel cycles; for example, normalized on the same electrical energy production, one has for the BWR-fuel: burnup 27.5 GW·day/t; enrichment 2.7%; total uranium load 130 t, of which one-fourth is renewed yearly; 280 kg extracted plutonium yearly, of which 190 kg fissile isotopes. For the CANDU-fuel one has 7.5 GW·day/t; natural uranium (no enrichment); whole uranium load of 130 t is renewed yearly; 350 kg extracted fissile plutonium 239 yearly. In all three cases the fuel is in the chemical form of uranium-oxide (UO_2); thus the fuel weight is 13% more than the pure uranium load (the data in Refs 3.2, 3.3, and 3.4 are mean values deduced from Refs. 3.5, 3.6, and 3.7).

3.3 *Consumed uranium.* Eighty percent of the consumed uranium 235 (i.e., 555 kg) is fissioned and the rest (135 kg), as a result of neutron capture, is transmuted into uranium 236 (125 kg) and into neptunium 237 (10 kg). Of the 650 kg of uranium 238 consumed, 45 kg are fissioned by fast neutrons and 600 kg are transmuted into plutonium (of which 530 kg into the fissile isotopes 239 and 241, and 70 kg into the nonfissile isotopes 240 and 242), and into americium 243 and curium 244 (less than 5 kg).

3.4 *Produced waste.* The 960 kg of fission waste result from the fission of 55 kg of uranium 235, 310 kg of plutonium 239, 50 kg of plutonium 241, and 45 kg of uranium 238. The nonfissile waste consists of 140 kg of various actinides (i.e., 125 kg of uranium 236, 10 kg of neptunium 237, less than 5 kg of americium 243 and curium 244), plus 10 kg of plutonium 242 and 60 kg of plutonium 240 (the latter, however, is easily transformed by neutron capture into the fissile plutonium 241). From a point of view of nuclear transmutations, one can say that the PWR yearly transforms 1340 − 125 = 1215 kg of uranium into 960 kg of intermediate elements (fission products) and into 255 kg of various trans-uranic elements, in addition to energy that is nearly equivalent to 1 kg of matter. The mean thermal energy per fission is taken to be 200 MeV (Ref. 3.5, p. 513), corresponding to 22.6 MWh/g of fissioned uranium or plutonium.

3.5 B. L. Cohen, *Reviews of Modern Physics 49*:1 (1977): High-level radioactive waste from light-water reactors. Also in *Scientific American 236*:10 (June 1977): The confinement of radioactive waste.

3.6 American Physical Society Report in *Reviews of Modern Physics 50* Nr. 1/II (January 1978): Report on nuclear fuel cycles and waste management.

3.7 F. David and J.-P. Schapira, *La Recherche 11*:520 (May 1980): Le retraitement des combustibles nucléaires.

3.8 *Isotope separation by centrifuges.* The separation of the isotopes is obtained in ultracentrifuges with the vertical rotor hold by magnetic pivot and bearing, and spinning with peripheral velocities of about 400 m/s corresponding to up to 100 000 rpm. To obtain reactor grade enrichment of 3%, about 10 stages in series (each equipped with many centrifuges) are required versus 1 200 stages in the diffusion process. The U.S. Department of Energy has estimated the cost of the centrifuge enrichment to be lower by 30% as compared with the one by diffusion; however, it must still be shown that the lifetime of the highly stressed centrifuge components will be long enough (more than 10 years) to make this estimate come true. Both the centrifuge and diffusion method operate with uranium hexafluoride (UF_6—usually called "hex"), the only suitable gaseous compound of uranium. It is highly corrosive and brings substantial handling problems. Further reading: S. Whitley, *Physics Technology* *10*:26 (1979): The uranium ultracentrifuge.

3.9   *Depleted uranium stocks.* It is instructive to roughly estimate the (partially depleted) uranium stock that has accumulated up to now as a consequence of nuclear energy exploitation both for civilian and military applications.

A 1-GWe light water reactor has more than 80 t of enriched uranium in its core, from the production of which about 500 t of depleted uranium have been derived. The preparation of the fuel loads of the light water reactors in operation or ordered until the beginning of 1982 with a power capacity of 450 GWe has produced about 230 000 t of depleted uranium; in addition, the total generation of about 3500 million MWh of electricity with these reactors implies an additional 80 000 t of depleted uranium. The preparation of a critical mass of 20 kg (Table 3.5) enriched at 93% with uranium 235 will produce an uranium stock of 4.1 t depleted to about a 0.25% content in uranium 235. If one assumes the existence of 20 000 charges with a mean critical mass of 20 kg of uranium each, one obtains a stock of about 100 000 t of depleted uranium. The total thus would amount to roughly 410 000 t.

In fact, it is reported that the United States alone have a depleted uranium stockpile of more than 200 000 t and Britain of 20 000 t (*Financial Times*, May 1, 1978).

3.10 *Genealogy of heavy nuclei.* The actinides (i.e., the chemically similar metallic elements with atomic number 89—actinium—and higher, thus including the transuranic elements) decay mainly by α-particle emission; for example, plutonium 239 decays with a mean half-life of 24 000 years into uranium 235, and then—after 710 million years—into thorium 231. The emission of particles— i.e., negative or positive electrons—is also possible, the result being a change of the atomic number at constant mass number; for example, thorium 231 decays—after 25 hours—by emitting a negative electron into protactinium 231 (with atomic number 91). The complex decaying schemes involving nearly 100 heavy isotopes follow a certain order, since the α-emitting nuclei can be classified into one of four decay series that are characterized by mass numbers with multiples of 4:

Thorium series: including, e.g., ^{240}Pu, ^{236}U, ^{232}Th, and ending at ^{208}Pb (lead)
Neptunium series: including, e.g., ^{241}Pu, ^{237}Np, ^{233}U, and ending at ^{209}Bi (bismuth)
Uranium series: including, e.g., ^{242}Pu, ^{238}U, ^{234}Th, and ending at ^{206}Pb
Actinium series: including, e.g., ^{243}Cm, ^{239}Pu, ^{235}U, and ending at ^{207}Pb

3.11 *Reprocessing of spent fuel.* The spent fuel rods from civilian reactors must first be cut mechanically into small pieces and these are then immersed in hot nitric acid to free the fuel from the alluminium or zirconium jackets. By applying the TBP solvent (tributyl phosphate), the chemical extraction proceeds in three successive steps: First to be extracted are the highly radioactive waste composed of the fission elements, and most of the transmutation elements, as americium, curium; then plutonium; and finally uranium. The TBP solvent is purified and reused.

Several plants are in operation around the world that have reprocessed up to now
10 000 tons of fuel for weapons grade plutonium production: for example, Savannah
River (South Carolina), Hanford (Washington), Rocky Flats (Idaho), Windscale
(United Kingdom), Marcoule and La Hague (France).

Various plants for reprocessing spent civilian fuel are designed or under construction
(capacities and dates of operation could still be subjected to substantial changes): In
La Hague a second unit with a capacity of 800 tons/year should come into operation
in 1988 and a third could follow later; at Sellafield a 1 200 ton/year plant is expected
to be operational in 1990; in Germany a sophisticated plant (1 400 tons/year) was
foreseen at Gorleben and is now given at Wackersdorf, but has not been commissioned yet; the United States has a new plant (1 500 tons/year) ready at Barnwell
(South Carolina), but due to the restraining nuclear policy of the late Carter admin-
istration and the lack of interest by private industry it has not yet (1985) been put
into operation.

3.12 *The Bulletin of the Atomic Scientist*, October 1979, p. 15; also *The Economist*,
March 14, 1981; *National Geographic*, April 1979, p. 477, specifies: "500 000
tons of highly radioactive material and 64 million cubic feet of less radioactive
trash".

3.13 *Reprocessing cost.* The extent of the financial commitment necessary to build
and run the reprocessing plants is reflected in the cost of $300 000/ton required
in the contracts for reprocessing irradiated nuclear fuel drawn up in 1977. (This
price does not include the cost of long-term storage.) Present (1982) estimates
seem to indicate that the unitary cost will be about twice the present one if
referred to the reprocessing in future new facilities. On the other hand, there is
still an abundant offer of uranium on the market (at $100/kg) so there is no
pressure yet to use it more efficiently by reprocessing. Nor is there a clear
financial interest, since with fissile plutonium prices at $30 000 to $40 000/kg the
bonus of extracting uranium and plutonium (see Fig. 3.9) amounts to approximately $100(U) + 200(Pu) = \$300/kg$ of spent fuel.

3.14 *Radioactive Waste Management*, Proceedings of an International Conference
held in Seattle, Washington, May 16–20, 1983 (Int. Atomic Energy Agency,
Vienna 1984): Vol. 1, Management Policy; Vol. 2, Handling; Vol. 3, Storage;
Vol. 4, Environment.

3.15 W. G. Thilly, *Technology Review*, p. 37 (February/March 1981): Chemicals,
genetic damage, and the search of truth. *Identifying and Estimating the Genetic
Impact of Chemical Mutagens* (National Academy Press, Washington, D.C.,
1983).

3.16 G. Greenhalgh, *The Necessity for Nuclear Power* (Graham and Trotman,
1980).

3.17 W. Marshall, *Physics Technology* 9:115 (1978): Nuclear power and the proliferation issue. A. B. Lovins, *Nature* 283:817 (1980): Nuclear weapons and
power-reactor plutonium.

3.18 Article VI of the *Non-Proliferation Treaty states:* "Each of the Parties of the
Treaty undertakes to pursue negotiations in good faith on effective measures
relating to the cessation of the nuclear arms race at an early date and to
nuclear disarmament, and on a treaty on general and complete disarmament
under strict and effective international control." As of the beginning of 1982,

116 countries had signed the treaty. Nonsigners were Albania, Algeria, Angola, Argentina, Bahrain, Bhutan, Brazil, Brunei, Burma, Cambodia, Chile, China, Cuba, Equatorial Guinea, France, Guinea, Guatemala, Guyana, India, Israel, Malawi, Mauritania, Monaco, Mozambique, Oman, Niger, North Korea, Pakistan, Papua New Guinea, Qatar, Sao Tome, and Principe, Saudi Arabia, Seychelles, South Africa, Spain, Tanzania, Uganda, United Arab Emirates, Vietnam, Zambia, Zimbabwe. (*Source:* Jozef Goldblat, Agreements for Arms Control, Stockholm International Peace Research Institute, 1982.)

3.19 J. J. MacKenzie, *Technology Review* (February 1984), p. 34: Finessing the risks of nuclear power. S. Levine, ibid., p. 40: Probabilistic risk assessment: Identifying the real risks of nuclear power.

Chapter 4

4.1 *Nuclear fission energy resources—Uranium.* Its total energy content, corresponding to 200 MeV per fission of one nucleus, is 82×10^{15} J/t or 78 Q per million tons of uranium. Even in an ideal breeding fuel cycle, only about 70% of the initial fuel can be fissioned (most of the rest being transformed to nonfissile actinides), and the energy content is thus reduced to 55 Q per million tons of uranium.

Uranium is contained in very diluted forms in different types of mineralization; prospecting has up to now generally been limited to deposits with a yield of more than 500 g uranium per ton of ore, whereas the average abundance in the earth's crust is less than 4 g per ton. The potentially available uranium resources increase as one considers the prospection of ore deposits with decreasing uranium content (thus increasing extraction costs). There is about a 300-fold increase in recoverable uranium for each tenfold decrease in ore grade with regard to presently exploiled ores. (K. S. Deffeys and I. D. MacGregor, *Scientific American*, January 1980: World uranium resources). Production in 1980 reached a height of 41 000 t of uranium.

In 1979 the Nuclear Energy Agency of OECD has estimated the world reserves of uranium in the price range up to \$130/kg of yellow cake (U_3O_8) at about 4.3 million tons (corresponding to an energy content of 230 Q). Relaxation from this price constraint has allowed the International Uranium Resourses Evaluation Program to estimate the world's total uranium potential at 26 million tons (1 400 Q). There is much more uranium in the earth, but at much lower concentrations which are not considered useful with foreseeable technologies, such as in marine shales, granite, and particularly in the seawater. In the latter, a remarkably constant content is found over the various oceans of 3.3 micrograms uranium per liter, for a total of 4 500 million tons (250 000 Q), as all the oceans have a water content of $1.37 \, 10^{21}$ liters.[4.24]

Thorium. Thorium also produces fission energy when inserted in a thermal breeding cycle and provides the same energy content as uranium, i.e., a practical 55 Q per million tons of thorium. Potentially interesting thorium resources are even less well known than uranium resources (uranium has attracted more attention because of its directly fissionable uranium 235 isotope content). As the average abundance in the earth's crust of thorium is 12 g/ton, i.e., three times more than uranium (*Handbook of Chemistry and Physics*, 37th ed.), the relevant energy reserves from thorium could be roughly estimated at three times more than those of uranium. However, there are 1971 estimates (J. P. Holdren, Report UCID-15953, Lawrence Livermore Laboratory, Livermore, 1971: Adequacy of lithium supplies as a fusion energy source) that indicate the U.S. thorium resources to be about 10 times those of uranium (30 versus 3.1 million tons), when related to a cost of not more than about $150/kg.

4.2 *Nuclear fusion energy resources—Lithium*. Lithium is used for breeding tritium, which is burned in the deuteron-triton fusion process, giving 17.6 MeV per reaction; this corresponds to 245×10^{15} J/t, or 230 Q per 1 million tons of lithium. Since blanket design of a pure fusion reactor requires in practice (at least) three times as many lithium 6 isotopes as contained in natural lithium, and since the energy production per fusion event is in this case $17.6 - 0.9 = 16.7$ MeV (including the energy loss due to the still prevailing lithium 7 reactions), the energy content is reduced to 73 Q per million tons of natural lithium. Notice that in the extreme case of using only lithium 6 for breeding (giving 22.4 MeV per fusion), the energy content reduces further to 22 Q per million tons of natural lithium.

At present lithium is extracted from pegmatite rocks or from the brine of salt lakes; it could also be easily obtained from geothermal and mineral springs, or from the water pumped from oil fields. The presently low annual world requirements of less than 7 000 t could increase in future in view of possible applications in energy technology, such as lithium sulfur batteries and particularly fusion power plants. This potential future interest has led in recent years to various appraisals of the lithium-supply situation. [For a discussion of lithium reserves, see R. Bünde, Report IPP 4/164 (Max-Planck-Institut für Plasmaphysik, Garching bei München, 1977): The energy reserves of DT-fusion]. Typically, the proven and probable reserves are put at 1.4 million tons, the possible reserves at 5.2, and the predicted resources at 4.3, for a total in these classes of 11 million tons of lithium (800 Q), a figure agreed by various authors [see, for example, H. Schmidt, Report of the Bundesanstalt für Bodenforschung (Hannover, September 1973): Lithiumlagerstätten in der Welt]. Estimates on total recoverable resources (including speculative resources) with high recovery costs, are approximately 140 million tons of lithium (10 000 Q).

Much larger resources of lithium are present in the oceans. As the lithium content in seawater is given by various authors between 0.1 and 0.19 mg/liter, the mean lithium content (corresponding to 0.145 mg/liter) in all oceans amounts to 2×10^{11} t (15 million Q).

Deuterium. The complete fusion of deuterons and their first generation reaction products ($6D \rightarrow 2\alpha + 2p + 2n + 43.3$ MeV) produces 350×10^{15} J/t or 330 Q per million tons of deuterium. As there is 1 deuterium atom per 6 700 hydrogen atoms in the earth's crust, the abundance is 0.42 mg/kg of crust material (the hydrogen content being 1.4 g/kg); and in the water the concentration is 33.3 mg/kg or liter of water, resulting in a deuterium content in all oceans of 4.6×10^{13} t (15×10^9 Q).

4.3 *Oil resources*. The energy content is (Table 1.1) 45.4×10^9 J/t or $43 \ 10^{-12}$ Q per ton of oil. Up to 1979 about 50 Gt (billion ton) of oil (2 Q) had been already produced and the world's current level of consumption is 3 Gt/year (0.12 Q/year). Estimates of reserves are continuously updated by various companies. According to the *Oil and Gas Journal*, on January 1981 the total identified reserves (i.e., proven, probable, and possible reserves of reasonable economic interest) were about 91 Gt (4 Q), of which 40% alone is located in the Arabian peninsula, and probably more than 10% in the Soviet Union. The estimated reserves have increased steadily in recent years, more than eight times since 1950!

As for nuclear energy, our interest here is for potential reserves available in the future, so as to complete in a comparable way Fig. 4.2. Of the many studies in this respect made in the last 30 years, we have taken the one presented by P. Desprairies at the World Energy Conference 1977 in Istanbul. It is based on a so-called Delphi poll, in which selected experts reply first to a questionnaire, and then, in a second round, review their opinions to establish the final result. The mean value of the majority of the replies given in 1977 concerning the remaining world recoverable resources of oil is around 300 Gt (14 Q). This figure includes oil from the ocean depths and polar zones, which is still regarded as unconventional, and assumes that the current rate of recovery of 25% will be increased to 40% within the next 20 years. Nevertheless, the result is probably more conservative than the one applied for nuclear resources, as it is limited to oil that can be extracted at production costs of less than twice the 1976 prices ($12 to $13 per barrel, in 1976 dollars). The upper, speculative resource included in Fig. 4.2 represents the most optimistic result to this inquiry: near to 1 000 Gt of oil (40 Q). It stands for a figure that implicitly relaxes the recovery rates and extraction costs assumed previously.

According to another source (AGIP, 1981), 140 Gt of potentially new resources could be added at the end of this century to today's estimated reserves of 90 Gt; 45% of the 140 Gt would result from new discoveries and the remaining 55% could be obtained through improved extraction techniques, inevitably at higher but still acceptable economic costs, from already known deposits. (The present standard extraction methods recover less than about one-third of the oil that originally lies in the deposits; see Ref. 4.26).

The survey of world resources, presented at the 1980 World Energy Conference in Munich, estimated the still available oil quantities at approximately 300 Gt, including those located in deep water and in Artic regions; 60% of these resources are not yet discovered, but are inferred from extrapolating informations regarding known reserves. An other source (B. Tissot, *La Recherche*, January 1982: Les nouveaux pétroles) gives somewhat higher figures for the ultimate resources: 300 to 350 Gt plus the deep water and Arctic deposits (for a total thus of about 400 Gt).

4.4 *Nonconventional oil resources*. Heavy oil, tar sands, and oil shales represent huge oil deposits from which, however, the extraction is difficult and expensive. Heavy oil, for example, is thick, viscous, and requires uncoventional extraction techniques. At present it is extracted in California and, on a small scale, in Venezuela. Famous heavy oil deposits are situated in the Orinoco belt in Venezuela. Venezuelan government officials estimated in 1979 that it should be possible to recover economically about 68 Gt of oil (3 Q) alone from these deposits. The heavy oil resources are estimated at up to 300 Gt, of which about one-third should be recoverable at prices within a factor of 2 of conventional oil recovery costs.

Tar sand deposits contain very thick oil, tar, or bitumen intermixed with sand. The tar sand deposits in Alberta (Canada), where commercial exploitation is only beginning, contain more than 90 Gt of oil, of which at least one-third should be recoverable.

Oil shales are the most important nonconventional oil deposits. They contain the solid kerogen, which produces a heavy oil by distillation; 30 to 250 liters per ton is thought to be an exploitable content, in future. Resources considered by the U.S. Geological Survey (1976) to be "near to possible commercialization" correspond to 27 Gt of oil, whereas "not yet economic resources" are around 440 Gt, in the ground. Of these, 300 Gt alone are in the Green River formation spreading over the states of Colorado, Utah and Wyoming, with a content of 120 liters/ton of shale. About 120 Gt are expected to lie in the Irati deposit in Brasil.

More than any other energy source, the knowledge of unconventional oil resources is scant. Up to now, these resources have never systematically been looked for, but have just been found by chance or by prospecting for other energy or mineral resources. We feel, therefore, that the given figures are rather conservative. In Fig. 4.2 we include, as the lower figure, the energy content of reserves with near to present recovery costs: 100 Gt from heavy oil + 30 Gt from tar + 27 from shale = 157 Gt of oil (7 Q). The larger figure includes, in addition, resources that are estimated to be not easily recoverable nor economic: 300 Gt from heavy oil + 90 Gt from tar + 440 Gt from shale = 830 Gt of oil (35 Q).

British Petroleum estimated in 1978 that hydrocarbons contained in oil shales correspond to 400 to 500 Gt of oil, and those in heavy oil and tar sands to 400 to 680 Gt, for a total of 800 to 1220 Gt of oil (35 to 52 Q) in nonconventional oil resources. But how much of this can be recovered? If we assume 20% for heavy oil and tar sand, and 10% for oil shale in the medium term and an overall 40% in the very long term, we obtain the two extreme figures of approximately 5 and 20 Q, which correspond to nearly half of the energy resources indicated in Fig. 4.2. Another detailed source (B. Tissot, *La Recherche*, January 1982: Les nouveaux pétroles) indicates the equivalent oil content of heavy oil and tar sands of 370 to 600 Gt and of oil shale of 500 Gt, similarly to the previous information.

4.5 *Natural gas resources*. For the mean energy content of natural gas at normal pressure and temperature we take 40 MJ/m^3, corresponding to 3.7 × 10^{-14} Q/m^3. According to a 1982 study by the International Energy Agency (IEA), the natural gas reserves are estimated at 77 × 10^{12} m^3 (3 Q), of which the Soviet Union owns more than one-third. Extrapolated resources contain an additional 190 × 10^{12} m^3, for a total of 10 Q. These figures are probably conservative, since continuously new reserves and resources are being added. According to another source (AGIP, 1981), the estimated and extrapolated resources amount in total to 207 × 10^{12} m^3 (8 Q).

Here we should recall the exciting but highly speculative hypothesis mentioned in Chap. 2 according to which huge methane resources of nonfossil origin could exist in the earth, possibly many orders of magnitude larger than the previously described ones.

4.6 *Coal resources*. Various types of coal (or solid fossil fuels) with different calorific values exist, such as anthracite, lignite, or brown coal. We reduce the different quantities in units of tons of coal equivalent (tce), such that the energy content is 2.93 × 10^{10} J/tce (i.e., 7000 kcal/kg), or 28 × 10^{-12} Q/tce.

At the recent World Energy Conference, a group of experts estimated the world resources at 10 000 Gtce (billion tons of coal equivalent), or 280 Q. On equal footing with the other information presented in Fig. 4.2, this figure must be considered a conservative estimate, since it includes only coal resources located within 1 200 m depth, and with little extrapolation. Of the 10 000 Gtce, nearly half (4 900 Gtce) is located within the Soviet Union, one-quarter in the United States, and one-seventh in China. In comparison, Europe is relatively poor in coal resources: the Federal Republic of Germany possesses about 250 Gtce, the United Kingdom 160 Gtce, and Poland 130 Gtce.

Technically and economically recoverable reserves are indicated at 6% of the mentioned resources. However, most experts consider this quantity so highly conservative that even doubling to 1 200 Gtce (3 Q) appears to be a sound estimate.

4.7   A. Rose, ORNL Report No 5506 (Oak Ridge National Laboratory, Oak Ridge, Tenn. 1979): Energy intensity and related parameters of selected transportation modes; Passenger movements. J. Meyers and L. Nakamura, *Saving Energy in Manufacturing: The Post Embargo Record* (Ballinger Books, Cambridge, Mass., 1978).

4.8   Gedanken zum Energiekonzept der Schweiz, Technische Rundschau Sulzer No 4, 1975.

4.9   *Breeder reactors.* The curves in the figure included in Ref. 3.1 show that uranium 235 can provide effective breeding only when absorbing neutrons with energies larger than 500 000 eV (electronvolt); uranium 233 for energies above 4 000 eV and marginally below 1 eV; and plutonium 239 for energies larger than 30 000 eV. These figures must be compared with the neutron energy spectrum provided by a real reactor assembly and lying anywhere in between the thermal and the fast reactor spectra. The curves suggest the possibility of supporting full breeding with an uranium-plutonium breeding cycle in a fast reactor where, with no moderator, a breeding factor of 1.1 to 1.5 can be obtained in practice (see the definition in the main text).

On the other hand, a light water reactor using enriched uranium achieves a breeding factor of about 0.5. Values in between are reached by various fuel combinations and different reactor concepts, which are called *converters* because they convert an appreciable part of fertile to fissile fuels without, however, attaining the self-sustainment factor of 1. For example, the uranium-thorium breeding cycle in a quasi-thermal reactor, by taking advantage of the uranium 233 breeding capability below 1 eV, can approach marginally this limit.

Plutonium 239 is the basic fissile material of a fast reactor where it is fissioned with high efficiency. This isotope could be also used in light water reactors but less efficiently; in fact here thermal neutrons cause 85% fission in uranium 235 versus only 65% in plutonium 239.

The data reproduced in Fig. 3.9 show that to generate heat, the modern light water reactor can consume only $0.035/(3.3/0.7) = 0.75\%$ of the uranium initially required (before enrichment); the percentage increases if uranium and plutonium are recovered by reprocessing and then reused. A well-balanced system of fast breeder reactors and reprocessing plants, on the other hand, could increase in the long term fuel exploitation by over 70 times, as it can consume ideally 50 to 70% of the uranium initially fed into the cycle. Exploitation of the uranium energy content cannot be larger because there are inevitable losses in the fuel cycle and particularly because a fraction of neutron captured in uranium and in the subsequent heavy isotopes lead to nonfissile elements. For example, the fissile fraction of plutonium is only 70% at its first cycle (Table 3.2), but after all the extracted plutonium is reinserted into the reactor it drops at every recycle and reaches 40% at the fourth recycle.

4.10 *Power balance on earth's surface.* Solar radiation is by far the largest power input. As the power emitted by the sun is 3.8×10^{26} W and the average distance between the sun and the earth is 150 million km, the average radiation density at the earth's position is 1.35 kW/m². On the earth this gives an average radiation of 1.35:4 = 0.34 kW/m², and the solar power reaching the earth's surface is thus on average $0.47 \times 0.34 \sim 0.170$ kW/m², as 30% is reflected back directly by the atmosphere and 23% is absorbed by evaporation. The average radius of the earth is 6370 km, so that the total solar power influx onto the earth is $174\,000 \times 10^{12}$ W (given as 100% in Fig. 4.1), of which 47%, or nearly 90000 TW, reaches its surface. This corresponds to 10000 times the power generated artifically by man (9 TW) derived at present nearly completely from the fossil and nuclear fuel stores. There are additional minor contributions from the earth's interior (from thermal, nuclear, and gravitational sources) and from outer space through cosmic radiation (particles and electromagnetic waves) and gravitation (tidal energy).

4.11 IIASA—International Institute for Applied Systems Analysis, Laxenburg, *Energy in a Finite World*, Vols. I and II (Ballinger Books, Cambridge, Mass. 1981).

4.12 W. E. J. Neal, *Physics Technology 12*:213 (1981): Thermal energy storage.

4.13 S. Wieder, *Introduction to Solar Energy for Scientists and Engineers* (Wiley, New York, 1982). D. K. McDaniels, *The Sun: Our Future Energy Source* (Wiley, New York). M. Iqbal, *An Introduction to Solar Radiation* (Academic Press, Toronto, 1983).

4.14 A. F. Beijdorff, *Energy Efficiency* (Shell Co., London, April 1979).

4.15 D. Merrick and R. Marshall, *Energy—Present and Future Options* (Wiley, New York, 1981). See also Ref. 1.9.

4.16 M. H. Ross and R. H. Williams, *Our Energy: Regaining Control* (McGraw-Hill, New York, 1981). National Audubon Society, *Audubon Energy Plan* (New York, 1981). G. Leach et al., *A Low Energy Strategy for the United Kingdom* (International Institute for Environment and Development, London, 1979).

4.17 C. L. Wilson, *Coal-Bridge to the Future: Report of the World Coal Study* (Ballinger Books, 1979). D. Pines et al., *Reviews of Modern Physics 53*, no. 4, part II (October 1981): Research planning for coal utilization and synthetic fuel production. Ch. Cunningham, *New Scientist*, p. 447 (February 18, 1982): Gas from coal that can't be mined.

4.18 P. Auer (ed.), *Advances in Energy Systems and Technology* Vol. *II* (Academic Press, New York 1979)—B. K. Hartline, Science *209*, 794 (1980): Tapping Sun-Warmed Ocean Water for Power.

4.19 J. C. Rowley, *Physics Today*, p. 36 (January 1977): Geothermal energy development. H. C. H. Armsteda, *Physics Technology 11*:2 (1980): Future prospects for geothermal energy.

4.20 P. N. Vosburgh, *Commercial Application of Wind Power* (Van Nostrand

Reinhold, New York, 1983).

4.21   A. L. Fahrenbruch and R. H. Bube, *Fundamentals of Solar Cells—Photovoltaic Solar Energy Conversion* (Academic Press, New York, 1983).

4.22   H. Munser, *Fernwärmeversorgung* (VEB Deutscher Verlag für Grundstoffin-dustrie, Leipzig 1980). E. Sprenger, *Taschenbuch für Heizung und Klimatechnik* (R. Oldenbourg, München 1981). H. Bachl, Fernwärme International—FWI *8*, Heft 3 (1979): Kann Fernwärme über Latentwärme wirtschaftlich übertragen werden?

4.23   J. R. Howell, R. B. Bannerot, and G. C. Vliet, *Solar-Thermal Energy Systems* (McGraw-Hill, New York, 1982).

4.24   K. Schwochau, in *Topics in Current Chemistry*, Vol. 124, p. 91 (Springer, Berlin 1984): Extraction of Metals from Seawater.

4.25   R. D. Heap, *Heat Pumps* (E. & F. N. Spon, London, 1981).

4.26   G. de Lamballerie, *La Recherche 12*:148 (1981): La récupération améliorée du pétrole. F. S. Ellers, *Scientific American*, p. 31 (April 1982): Advanced offshore oil platforms.

4.27   M. M. El-Wakil, *Power Plant Technology* (McGraw-Hill, New York, 1984).

4.28   *Energy from Biomass*, Proceedings of the Third International Conference, Venice, March 25–29, 1985 (Commission of the European Community, Luxembourg, 1985). See also Ref. 5.22.

4.29   F. A. Curtis (ed.), *Energy Developments: New Forms, Renewables, Conservation* (Pergamon Press, Toronto, 1984).

4.30   Information about the broad based program on renewable energy sources, energy conservation, and thermonuclear fusion, sponsored or coordinated by the European Community, can be obtained from the Directorate General XII, Science Research and Development, 200 Rue de la Loi, Brussels.

Chapter 5

5.1 *Fusion reactions*. There exists a large number of exothermic fusion reactions, even if only the isotopes of the five lightest elements—hydrogen, helium, lithium, beryllium, and boron—are considered. The most important reactions are described in the following table, whereby it may also be interesting to recall the natural abundances in % of the isotopes: $p(99.985)$, $D(0.015)$, $T(\sim 0)$; $^4He(100)$; $^6Li(7.4)$; $^7Li(92.6)$; $^9Be(100)$; $^{10}B(18.83)$, $^{11}B(81.17)$.

n	Reaction (Energy Fractions in MeV)	Total Energy Release (MeV)
	Fusion reactions	
1a	$D + D \rightarrow {}^3He(0.82) + n(2.45)$ (50%)	3.27
1b	$D + D \rightarrow T(1.00) + p(3.03)$ (50%)	4.0
2	$D + T \rightarrow {}^4He(3.52) + n(14.08)$	17.6
3	$D + {}^3He \rightarrow {}^4He(3.7) + p(14.7)$	18.4
4	$p + {}^{11}B \rightarrow 3{}^4He$ (2.89)	8.67
5	$p + {}^6Li \rightarrow {}^3He(2.30) + {}^4He(1.72)$	4.02
6	${}^3He + {}^6Li \rightarrow p(12.40) + 2\ {}^4He$ (2.25)	16.9
7	$p + {}^9Be \rightarrow {}^4He(1.38) + {}^6Li(0.85)$	2.13
	Triton breeding	
8	$n + {}^6Li \rightarrow {}^4He + T$	4.8
9	$n + {}^7Li \rightarrow {}^4He + T + n'$	−2.87

A fusion reaction is, in general, more difficult to achieve than a fission reaction. Setting up fusion contact between two nuclei—both electrically charged—involves overcoming their coulomb (or electrostatic) repulsion; the neutron which provokes fission, on the other hand, penetrates the nucleus virtually unhampered. The deuteron reactions with the equally probable reaction channels n. 1a and 1b have the merit of relying only on the readily available nonradioactive deuterium fuel, thus excluding the complex tritium breeding and management problems; however, the probability for deuteron fusion is about 100 times smaller than for deuteron-triton fusion. Concerning the energy release of deuteron fusion, one should also add the reaction energies of the first generation triton and helium 3 products that could, in principle, be burned up nearly as fast as they are produced. The net result of adding up reactions 1, 2, and 3 gives then $6D = 2({}^4He + p + n) + 43.3$ MeV, i.e., a fusion energy release of 3.6 MeV per nucleon entering into the reaction, as compared with 0.85 MeV per nucleon in fission (1 MeV = 1 million electronvolt energy; see Table 1.1).

For the long term, basic interest can be paid to fuel cycles that produce no, or relatively few, neutrons, because they would reduce to a minimum the radioactivity of the reactor structures induced by neutrons. An example of a relatively neutron poor fuel is based on the deuteron + helium 3 reaction n. 3 that produces only charged nuclei although in the reacting mixture there would be the neutrons from the competing deuteron reaction n. 1a.

Scenarios have been developed, in which the relatively clean and safe helium 3 based reactors are operated near the consumers, whereas the nonradioactive helium 3 fuel is produced through reaction 1a in fusion reactors that can be located in remote areas.

Of distinct interest are the truly neutron-free fuel cycles because: (1) no thick and sophisticated blanket is needed; (2) radioactivity and radiation damage is at a minimum; (3) direct energy conversion—from the charged reaction products to electrical power through a magneto-hydrodynamic process (which might provide 60 to 70% efficiency)—is most convenient; and (4) the reaction energy contributes fully to the self-sustainment of the plasma burning. Among them, the most interesting is the proton-boron fuel based on reaction 4, also because boron content in the seawater assures resources of typically 100 million Q. The resources of the proton-lithium 6 fuel (reaction 5) are those of lithium 6 (see Ref. 4.2); if one considers only the reaction energy of 4.02 MeV one gets less than 1 million Q. The proton-beryllium 9 reaction 7 is less interesting, as the limited beryllium resources do not allow it to be an important fusion fuel. Unfortunately, ignition of these three particular thermonuclear fusion fuels can be reached only very marginally, or not at all, even with the most optimistic assumptions.

    5.2   *Tritium breeding.* The primary elements for the deuteron-triton fusion process consists of deuterium and lithium. As the latter is the less abundant of the two, it is the availability and breeding efficiency of lithium that determines the overall energy resources of this fuel cycle.

The breeding process is quite different in the isotopes lithium 6 and 7, contained in natural lithium with the abundances of 7.4 and 92.6%. Neutrons with energy smaller than 4 MeV can practically only induce the exothermic lithium 6 reaction 8 in Ref. 5.1 and are thereby lost. For neutrons above 5 MeV the endothermic lithium 7 reaction 9 strongly predominates, from which the neutron reemerges with lesser energy. For example, a neutron of 14 MeV from the deuteron-triton reaction entering into a slab of pure natural lithium induces with high probability the lithium 7 reaction, and the secondary neutron then enters in the lithium 6 reaction, with the overal result of consuming one each of the isotopes and producing about 1.9 tritium atoms and $14 + 4.8 - 2.8 = 16$ MeV of energy. Reactor blanket concepts generally include, in addition to the lithium isotopes and structural material, neutron multiplying elements such as beryllium, lead, uranium, or thorium. In these elements the absorption of a 14-MeV neutron is followed by the emission of two, sometimes three, neutrons of lesser energy that can breed tritium in the lithium 6 isotopes.

In conclusion, fusion reactor blankets can be designed in many different ways to assure the necessary tritium supply. However, in all practical cases, the relative lithium 6 over lithium 7 consumption is more than three times its relative content in natural lithium.

    5.3   *Nonthermonuclear fusion.* The fusion reaction can be achieved in the laboratory by shooting one nucleus against another; for example, by accelerating the deuterons to high velocities in an accelerator and then having them collide with a solid target composed of tritium atoms. A first loss of the projectile's energy is caused by detaching the electrons from their atomic bound (to which they nearly immediately return, irradiating the binding energy); but even if the target atoms were in the fully ionized plasma state, more energy is spent to accelerate the particles than energy is gained from the rare fusion events they cause.

In fact, at the particles energies of interest, the probability for the approaching projectile-target nuclei of coming within the range of the attracting nuclear forces, at about 10^{-15} m, and of undergoing a fusion reaction is much less than of being just scattered elastically (30 times less at 100 keV kinetic energy, or about 100 million times less at 10 keV). As a consequence, a fusion source will almost inevitably be a thermonuclear process, since a nucleus will, on the average, be randomized by many elastic collisions with other particles before it induces a fusion reaction.

By the way, the reaction probability of the easiest fusion process—the one involving deuterium and tritium—is at its maximum, at 100-keV bombarding energy, 100 times smaller than the fission by a thermal neutron of uranium 235, and it is more than 1 million times smaller at 10-keV bombarding energy. The relatively low-fusion probability (which makes this nuclear energy process so more difficult to be exploited on a practical level) is mainly a consequence of the fact that nuclei of similar electrical charges must be brought close together against the coulomb repulsion.

5.4 J. Jensen, *Energy Storage* (Newnes-Butterworths, London, 1983). Also Ref. 5.6.

5.5 R. L. Cohen and J. H. Wernick, *Science 214*:1081 (1981): Hydrogen storage materials, properties and possibilities.

5.6 I. Glendenning, *Physics Technology 12*:103 (1981): Compressed air storage.

5.7 Ch. A. McAuliffe, *Hydrogen and Energy* (McMillan, London 1980). L. O. Williams, *Hydrogen Power* (Pergamon Press, Oxford, 1980).

5.8 G. Genta, *Kinetic Energy Storage: Theory and Practice of Advanced Flywheel Systems* (Butterworths, London 1985).

5.9 D. P. Snowden, *Scientific American*, p. 84 (April 1972): Super-conductors for power transmission.

5.10 Cl. Gary, *La Recherche 10*:222 (March 1979): Le transport de l'énergie électrique.

5.11 Colin A. Vincent, *Modern Batteries: An Introduction to Electrochemical Power Sources* (Arnold, London, 1984). See also Ref. 5.4 and 5.24.

5.12 H. Kahn and J. Simon, *The Resourceful Earth* (Basil Blackwell, London, 1984). (A confutation of the "Global 2000 Report"; see Ref. 5.15).

5.13 D. Wilson, *European Experiment* (A. Hilger, Bristol, 1981).

5.14 J. J. Duderstadt and G. A. Moses, *Inertial Confinement Fusion* (Wiley, New York, 1982).

5.15 *World Energy Outlook* (Exxon Corp., New York, December 1979). C. L. Wilson, *Energy: Global Prospects 1985–2000: Report of the Workshop on Alternative Energy Strategies (WAES)* (McGraw-Hill, Maidenhead, 1977). *World Energy Resources 1985–2020* (IPC Science and Technology Press, Guildford, 1978). Commission on Nuclear and Alternative En. Systems, CONAES, *U.S. Energy Supply Prospects to 2010* (Natl. Academy of Sciences, Washington, D.C., 1979). *Global 2000 Report to the President* (Government Printing Office, Washington, D.C., July 1980).

5.16 J. O'M. Bockris, *Energy Options: Real Economics and the Solar-Hydrogen System* (Taylor & Francis Ltd., London, 1980).

5.17 UN Statistical Office, *World Energy Supplies* (United Nations, New York, annually).

5.18 UN Statistical Office, *Statistical Yearbook* (United Nations, New York, annually).

5.19 S. Stucki, *Europhysics News 12*, Nr. 8/9, p. 9 (1981): Hydrogen production by water electrolysis.

5.20 E. M. Goodger, *Alternative Fuels-Chemical Energy Resources* (MacMillan, London, 1980).

5.21 A. P. Fickett, *Scientific American 54* (December 1978): Fuel-cell power plants.

5.22 K. V. Sarkanen and D. A. Tillman (eds.), *Progress in Biomass Conversion*, *Vol. I* (Academic Press, New York 1979). D. Pimentel et al., *Science 212*:1110 (1981): Biomass energy from crop and forest residues. Ph. Chartier et S. Mériaux, *La Recherche 11:* 766 (July/August 1980): L'énergie de la biomasse.

5.23 R. J. De Bellis and Z. A. Sabri, Special Report EPRI ER-510-SR (Electric Power Res. Inst., Palo Alto, Calif., June, 1977): Fusion power, status and options. B. Brandt, H. Th. Klippel, and W. Schuurman, Rijnhuizen Report 81-129 (FOM-Institut, Nieuwegein, Netherland, February 1981): On fusion and fission breeder reactors; The IIASA Report RR-77-8 reviewed and updated.

5.24 J. P. Gabano (ed.), *Lithium Batteries* (Academic Press, London, 1983)

5.25 P. F. Chapman and F. Roberts, *Metal Resources and Energy* (Butterworths, London, 1983).

5.26 B. J. Skinner, *Proceedings National Academy of Sciences* (US) 76 (9):4212 (1979): Earth resources.

5.27 E. Teller (ed.), *Fusion*, Magnetic Confinement, Vol. 1, A and B (Academic Press, New York, 1981).

5.28 H. Knoepfel and E. Sindoni (eds.), *Heating in Toroidal Plasmas*, Vols. 1 and 2 (Franchi, Città di Castello, Perugia, 1984).

5.29 G. Casini (ed.), *Engineering Aspects of Thermonuclear Fusion Reactors* (Harwood Academic Publishers, New York, 1981).

5.30 S. Jones, in *Proceedings of the Workshop on Muon-Catalyzed Fusion* (Jackson, Wyoming, 1984). Further references, e.g., in A. A. Harms, *Journal of Fusion Energy 3*:303 (1983): The equilibrium μ-D-T fusion cycle.

Subject Index

173